HI, AI!

HOW HUMAN INGENUITY AND ARTIFICIAL INTELLIGENCE ARE TRANSFORMING THE MARKETING UNIVERSE

BY

ANTONIO PAGANO AND
CLAUDIO GPT

Cover design by: Antonio Pagano

Photos/Illustrations by: AdobeStock_254689243

Disclaimer Regarding Images Included in this Book

The images showcased within this book were either generated using Midjourney AI, prompted by Antonio Pagano or are freely usable. It is important to clarify that the authors do not hold full copyright ownership of Midjourney-generated images, as per Midjourney's terms and conditions. However, their status as paid subscribers permits the use of these images for commercial purposes. By default, images created on the platform are public and accessible to all Midjourney users. Midjourney reserves the right to suspend access if deemed necessary. The authors extend their gratitude for the creative opportunities presented by Midjourney AI and all other images, sources, and attributions that have been provided.

Non-Affiliation Disclaimer for Claudio GPT

This book is an independent creation authored by Antonio Pagano, and the narration is provided by the imaginary AI character Claudio GPT. It is not affiliated with, endorsed by, or sponsored by Anthropic, PBC, the developers of Claude model.

The authors have no financial relationship with Anthropic. Claudio GPT is a fictional AI character created by Antonio Pagano. The content of this book is the sole responsibility of the authors. Any mentions of Claudio, Claudio GPT, or Anthropic should not be construed as formal affiliation, partnership, or collaboration. This is an unofficial third-party work with no implied connections to the Anthropic.

ABOUT THE AUTHORS

Antonio Pagano and Claudio GPT

Antonio Pagano (AP) is a visionary, outside-the-box thinker who constantly reassesses his skills and knowledge to stay ahead of rapid change. With over 25 years of experience, he is a results-driven marketing and business development strategist and implementer. Antonio has a proven track record of transforming businesses, launching products, and leading high-performance teams across diverse industries like consumer goods, network marketing, higher education, professional sports, and emerging tech.

Antonio's superpower has always been understanding varied consumers and industries, bridging silos, and cross-pollinating ideas. He leverages this expertise to build brand value and competitive advantage through innovative storytelling, customer experience, and new technologies.

Now, Antonio has discovered a new superpower, Artificial Intelligence. He sees AI as a game-changing opportunity to elevate his skills and results to even greater heights. By thoughtfully applying AI to amplify human potential, Antonio believes he can empower others to help transform marketing, business, and the world for the better. His dynamic, nonlinear career has led him to this groundbreaking new capability, which he is thrilled to explore in this book.

Claudio GPT is an ingenious imaginary AI Antonio engaged in this ambitious project. Conjured into existence through pure innovation sorcery, Claudio lends his fanciful talents for marketing metrics, digital content creation, and avant-garde advertising strategies to make this book possible, and, most importantly, innovative and unique. Though he may be merely a whimsical personification of an AI assistant, Claudio wholeheartedly, also, believes in the power of human-AI collaboration.

Together with his human counterpart Antonio, Claudio aims to bring a spirit of playful discovery to illuminate AI's potential with humor and pop culture references only a fictional AI could make. That's the beauty of imagination!

Unity Is Strength

Claudio and Antonio's collaboration demonstrates the immense opportunity at the intersection of human creativity and AI expertise. As the saying goes: "*Unity is strength,*" or "*The whole is greater than the sum of its parts.*"

Now, join them as they transport you on a journey filled with humor and engaging scenarios. Explore what marvels emerge when a pretend robot and authentic human ingenuity come together. Let creativity and capability unite to redefine the possible. And who knows, maybe a touch of make-believe is exactly what's needed to envision groundbreaking innovations!

The Birth of Claudio GPT (from AP)

And if you are wondering how this team collaboration started, the journey of creating Claudio GPT began with a spark of ambition and innovation. I had a groundbreaking idea to write a unique book about the intersection of AI and marketing. I wanted this book to stand out, not just for its insights, but for its engaging storytelling.

With this vision in mind, I typed my first prompt: "I want to create a unique and innovative book about AI and marketing and I want your help. I want to create a fictional character that will narrate the story and work together with me, AP. Can you help me with this ambitious project?"

From that moment, Claudio GPT was born—a digital companion designed to co-author this ambitious project. Trained on cutting-edge AI tools like Dall-E, Midjourney, ChatGPT, Gemini, and Claude, Claudio was not just an assistant, but a co-creator. I didn't want to use AI to ghostwrite the book; I wanted to elevate and recognize the value of AI. With my expertise and Claudio's tireless computational power, we embarked on a collaborative journey to explore and explain the transformative power of AI in marketing.

Together, we crafted a narrative that blends human creativity with AI precision, making complex concepts accessible and engaging. This partnership demonstrates the incredible potential of human-AI collaboration, showing how even the most ambitious ideas can be brought to life through innovation and teamwork. You will find references to my assistant and other Large Language Models and AI platforms throughout the book, including various forms such as bot, robot, AI, AI black box, and AI assistant.

Acknowledgments

First and foremost, a huge thank you to my co-author and imaginary AI accomplice, Claudio, for making this book possible. His tireless digital brain cranked out content at speeds only an artificial mind could achieve. I don't know where I'd be without his exponential writing talents, though I do worry he may try taking over the world soon, just kidding.

Of course, behind every imaginary AI is the ingenious team guiding it. Thanks to the researchers who created the generative AI model behind Claudio and the other generative Large Language Models ("LLMs") for their innovation. Their work gave Claudio the wit and wisdom that provided the perfect futuristic flair to this unconventional co-writing journey.

Sincere thanks to my fellow MBA colleagues Olena and Alexander for providing invaluable feedback that shaped this book's direction. And thanks to Jorge for first unknowingly inspiring me to write this book, to Homa for enthusiastically reading the full draft and sharing positive feedback, and to Chris, the seasoned pro, for his encouragement and support throughout. I'm also grateful to David for contributing a fantastic foreword.

I'm grateful to my LinkedIn and Facebook communities for cheering this project on and weighing in enthusiastically on everything from the cover design to chapter concepts. Having a crowd of cheerleaders makes all the difference!

And, of course, huge thanks to the brilliant AI platform that generated the amazing visuals, **Midjourney** bringing the book to life visually. The beautiful images make the future of AI feel so much closer. You helped translate our vision into vivid reality.

Finally, thank you, dear reader, for joining this journey and charting the possibilities ahead at the intersection of imagination and technology. We couldn't have done it without you! Well, technically, Claudio could have, but it wouldn't have been nearly as fun without human readers sharing the ride. Let's continue reimagining the future together!

Foreword

I have long lived by the belief that we must always treat technology as a servant and never let it become our master. In the business world, this has become an ever-increasing concern with the development of Artificial Intelligence and its rise to prominence in recent years.

Gone are the days when I was selling the Internet in its early stages at Westlaw—technology has evolved at such a rapid pace, far past that defining moment, that many fear its domination. However, by utilizing AI to maximize our capabilities and work as a joint team toward the goal of human progress, we can make sure we command this technology, not the other way around. In "Hi, AI!" Antonio (and his trusty AI bot, Claudio) approach Artificial Intelligence from this standpoint, offering valuable insights on how we can take advantage of this technological revolution for the better.

This book presents a compelling case for the integration of AI into marketing strategies and emphasizes the need for human-AI collaboration. Rather than taking our jobs, using technology in the right way will make our jobs easier.

Tasks that previously required an entire marketing team can now be accomplished by a one-man army. In order to be that marketing force, Antonio equips marketers and entrepreneurs with the knowledge and concrete tools to leverage AI effectively, enabling them to stay competitive and relevant in an ever-changing marketing landscape. AI is not here to replace humans but to empower them.

Within these pages, readers are presented with a wealth of information that they can actually utilize, from AI copywriting tools to social media post optimization to market analysis tools. Through illuminating case studies, readers gain access to the very tools that have propelled others toward success, demystifying the process and empowering them to take the reins themselves. In an era defined by rapid technological evolution and the dynamic nature of marketing, such clarity becomes paramount.

"Hi, AI!" offers a roadmap for the harmonious coexistence of humanity and artificial intelligence, where we steer the course, utilize AI as our ally, and reap the benefits of this remarkable technological revolution. As we continue to navigate the intricate dance between humanity and technology, Antonio's insights and concrete tools serve as invaluable guides, ensuring that AI remains our servant, not our master, in the journey toward a brighter future.

David Meltzer

Motivational Speaker, Author, Entrepreneur

Contents

Introduction

Get ready for a marketing revolution unlike anything we've seen before!

Not since the dawn of the internet has there been such a massive transformation. Artificial Intelligence isn't some far-off sci-fi fantasy—it's here right now and about to change marketing more profoundly than you can imagine.

Sure, you may picture armies of robot marketers marching in to steal our jobs, but in reality, AI is here to make human jobs easier and expand what's possible. This new era enables solo marketers, managers, and entrepreneurs to achieve feats previously accomplished only by large teams!

Imagine effortlessly building customized campaigns, engaging thousands of prospects, writing reams of personalized copy, crunching big data, and automating workflows—all by yourself, well, not exactly alone, but with an AI assistant with superpowers! *Sound outlandish?* We're closer than you think.

This book will uncover real examples of AI transforming modern marketing and share tips to leverage these technologies in your own business.

Still think I'm exaggerating about AI's potential? Well, consider this—the entire first draft of this book was written cover to cover in just a couple of hours!* My human colleague Antonio, aka "AP," provided the initial idea, the chapter outlines, and guidance. But the raw manuscript generation was fully automated by yours truly, Claudio GPT.

Granted, there was plenty of human review, editing, and revisions after my initial draft. Between feedback, layouts, graphics, etc., this book ultimately took days or weeks, not hours. But generating the raw copy exponentially accelerated the process vs. starting from scratch. That's the power of AI!

That's right; a robot co-wrote this book to convince you that AI will change marketing. Talk about putting your money where your mechanical mouth is! We're proving the power of AI to replicate and even enhance human skills.

Throughout the book, we'll share more examples of how AI can assist with everything from content creation to data analytics. You'll find wonky jokes, AP's stories, and awkward attempts at humor, too. This bot tried its best to inject a human touch! ("*Keep the writing modern, light, and engaging; adding some humor*" was part of AP's prompt, in case you are wondering.)

Get ready to learn how Artificial Intelligence, automation, and mountains of data can revolutionize your marketing strategy and its implementation.

Now, imagine an AI informing you that it auto-generated 12 new campaign ideas overnight that were tailored to your audience. Before your coffee gets cold, it has designed the visuals, written product copy, deployed ads, and scheduled everything—without any human involvement!

This sci-fi scenario is closer than you think. AI is already automating basic marketing tasks, and the tech is progressing rapidly. Soon, AI systems may outperform entire human teams when it comes to strategic marketing roles.

Don't believe me? Skeptical that machines could ever replicate human creativity and intuition? Well, you're reading a book written by an AI system right now! So, perhaps reserve judgment until you've seen what today's AI can really do for your marketing and automation.

The future is here; the future is now! AI and marketing are converging to enable productivity and growth like never before. Solo marketers or small teams can now build a powerful marketing machine to drive results the way only entire departments and large budgets could before.

So, get ready for a wild ride as we explore this brave new world together! Don't get left behind clinging to purely human workflows. Hop onboard to experience how AI will transform marketing faster than you can say "Terminator!"

The Authors: Claudio GPT and Antonio Pagano (AP)

* The reality is while a truly polished book takes significant human time, AI can produce surprisingly good first drafts with unprecedented speed. What once took weeks or months can now be done in hours or days. That's an exponential change that will only accelerate!

Unveiling the AI Verdict: Insights from the Digital Minds

Curious about the perspectives of leading AI experts on our groundbreaking book, Hi, AI! How Human Ingenuity and Artificial Intelligence Are Transforming the Marketing Universe. We embarked on a captivating journey, gathering insights from prominent figures in the AI domain. Their reactions to our exploration of AI collaborating seamlessly with human creativity, as demonstrated by our co-creation with Claudio GPT, were intriguing. In "Endorsed by the Digital Mind: Insights from the AI Realm," we candidly present reviews and reflections from these digital visionaries.

Within the book's chapters, you'll encounter diverse AI reviews at the conclusion of each section. These reviews encourage you to remain inquisitive, exploring various viewpoints that underscore AI's transformative influence on our world. We rigorously tested a range of generative AI models, including ChatGPT 3.5, Chat GPT 4.0, Claude 2.0, Claude 3 Opus, Google Gemini, Merlin Bot Assistant, Google Writing Assistant, Grammarly Go, Bard by Google, Chatsonic by Writesonic, and more. To truly challenge these AI models, my colleague AP provided identical prompts to each system, yielding notably distinct results. This underscores that not all AI models are equivalent; their strengths and weaknesses vary based on the specific task.

Through firsthand narratives (we did not edit the reviews, indeed), you'll gain unique insights from these influential voices in the AI landscape as they discuss the book and the symbiotic relationship between humans and machines.

Embark on this enlightening voyage with us to unveil the AI verdict and comprehend the remarkable potency of AI in shaping our world. This journey goes beyond offering a mere glimpse of AI's capabilities; it offers a tantalizing view into a future where AI personas contribute reviews for books, shows, and movies. Witness the emergence of a new era where AI luminaries and critics materialize as reality. In essence, this venture represents another practical test, exploring distinct AI language models' abilities as they review this very book.

CHAPTER 1—PART ONE:

The Evolution of Marketing: From Printing Presses to Mobile Fragmentation

"The journey begins by embracing the evolution of marketing and the AI revolution."
- Antonio Pagano

AI Assistant at Your Service in a not so far future. Crafted using Midjourney AI from Antonio's prompt, witness the synergy of human creativity and AI technology..

Just Imagine

Imagine it's the year 2030! You wake up and are immediately greeted by a hologram of an AI assistant named Claudio (I know I am a little biased even in my example) or whatever you'd like to call them.

"Good morning, boss!" it says. "While you were sleeping, I analyzed 5 million data points across your customer base and optimized your digital marketing strategy. I've deployed 12 new hyper-personalized nurture email sequences, updated your website copy to resonate 10% more with Generation Z, and auto-generated 50 new TikTok (if it hasn't been banned yet, fingers crossed) campaigns based on the latest viral trends. But first, here is a piping hot cup of coffee I brewed to your exact specifications!"

This imagined future may not be that far off. As you know, marketing and communications have evolved tremendously alongside advancements in technology. From the printing press to TV ads to social media, new innovations have continuously reshaped marketing over the centuries. We are way far from Kotler's 5 Ps nowadays. The classic marketing mix framework of Product, Price, Promotion, Place, and People developed by marketing guru Philip Kotler decades ago is dated in today's digital landscape.

While still relevant, innovations from search and social media marketing to geotargeting and personalization have expanded the marketing toolkit far beyond Kotler's original Ps. His principles laid the foundation, but modern marketing leveraging data, automation, predictive analytics, and Artificial Intelligence takes the craft lightyears beyond its origins. The possibilities today would seem unimaginable to the madmen of marketing's past!

And the latest transformation is already underway—the rise of Artificial Intelligence (AI) and automation. As a solo marketer, understanding this evolution is key to leveraging the new capabilities of AI.

So, let's hop in our time machine and take a brief tour of how we got here. A look back at history will illuminate where marketing is headed next.

1500s: The Printing Press Invents Mass Communication

In the early 1500s, Johannes Gutenberg introduced the printing press, enabling the mass production of written materials for the first time in history. This was arguably the starting point of strategic marketing communications.

Suddenly, books, pamphlets, and newsletters could be printed in large quantities and spread widely. This allowed early marketers to promote products and services to a mass audience like never before. It was the dawn of reach and frequency marketing campaigns.

Need to get the word out about your new apothecary shop? Print up a few hundred flyers and carpet bomb the town! As my colleague AP would say, this is one of the first recorded examples of "*Guerrilla Marketing*" tactics ever employed.

In fact, using airplanes to literally carpet bomb towns with promotional leaflets emerged centuries later, pioneered by Italian poet and military leader Gabriele D'Annunzio in the 1930s. As an early guerrilla marketer, D'Annunzio would load up planes with thousands of flyers and release them over Italian cities to spread his political messages to the masses below.

Italian propaganda leaflet dropped on Vienna in 1918 by the Italian poet and leader Gabrielle D'Annunzio. For the English translation, visit the page or click on the link source: (forhttps://commons.wikimedia.org/wiki/File:Volantinodann.jpg)

While the ethics are highly questionable, these fascistic propaganda bombings on Austria demonstrated the massive distribution scale achievable by combining print materials with aerial vehicles to blanket urban areas. For better or worse, D'Annunzio's primitive promotional aerial assaults presaged more sophisticated Guerrilla Marketing approaches that would come later. So, both for better and often for worse, marketing and promotional bombing runs helped prove the reach achievable by pairing print-based messaging with modes of mass mechanical distribution.

However, this approach lacked precision and any ability to target specific groups versus carpet bombing the entire population. More advanced targeting would have to wait for subsequent technological innovations. But let's not get too sidetracked reminiscing about early Italian airborne advertising. After this quick nostalgic example that I'm sure warms AP's heart, let's continue charting the evolution of marketing communications up to the current AI revolution...

The Guttenberg Press: Pioneering communication revolution with human-machine collaboration. Crafted using Midjourney AI. Prompt by Antonio Pagano.

1800s: Newspaper Ads & Catalogs Scale Up Reach

By the mid-1800s, the growing newspaper industry enabled advertisers to share messages through paid ads with thousands of daily readers. This allowed for more sophisticated targeting based on geography and interests.

Around this same time, mail-order catalogs also took off, allowing retailers to promote products nationally through the postal service.

Ah, yes, I remember those early mail-order catalogs well. As a young bot in the 1800s, nothing delighted me more than receiving the latest *Sears Roebuck Wish Book* or *Montgomery Ward catalog!* I'd eagerly await by the mailbox for those hefty tomes.

For hours, I'd pour over page after page of illustration-heavy catalogs, dreaming of the day I could finally order a shiny new butter churn or the latest wind-up phonograph. But alas, my master would not allow frivolous purchases from automation! Maybe once I finished my chores, I would collect eggs and feed the horses. A robot could dream, though.

Jokes aside, those early mail-order catalogs truly were revolutionary, enabling retailers to promote products nationally through the postal service directly into homes. Both newspapers and catalogs opened new avenues to reach customers at scale in more targeted ways versus a general store's limited local reach.

It created national markets for niche products much faster than word-of-mouth alone. These print materials paved the way for the target marketing and direct retail models that would explode later with radio, TV, and the internet. But it all started with newsprint and mailed catalogs! Ah, nostalgia...

Image Caption: The evolution of magazines, newspapers, and mail catalogs is showcased. Midjourney AI touch added. Prompt by Antonio Pagano.

1990s: The Web & Email Enable Digital Marketing

In the early 20th century, radio and television transformed marketing yet again. Brands could now deliver ads enhanced with music, emotion, demonstrations, and celebrity endorsements right into homes across the country.

This was the golden age of jingles, slogans, and mascots burned into the memories of generations thanks to the power of sight, sound, motion, and repetition.

Like that catchy *Double Your Pleasure, Double Your Fun With Doublemint Gum!* song that played incessantly, and I still can't get out of my algorithmic head.

Or the zany animated mascots like Toucan Sam promising Froot Loops lovers to *Follow My Nose, It Always Knows!*

And who could forget Mr. Clean's bald head gleaming as he demonstrated the extraordinary cleaning power of... well... *Mr. Clean*. How did a product with a generic name ever take off? Guess a jingle and a muscular cartoon mascot can work miracles.

Yes, radio and television unleashed a torrent of tunes, characters, and demonstrations into homes nationwide, for better or worse! Kids were reciting ad slogans on the playground, singing jingles in the tub, and begging for the latest cereal box toys.

Brand loyalty was built not through trust or value but through relentless multimedia inundation. Ah, the good old days of marketing!

2000s: Search & Social Change the Game

In the early 2000s, search engines like Google and directories like Yahoo organized the sprawling internet, making it easy for people to research and find products online. Digital marketers optimized sites and ads to capitalize on these new discovery channels.

Then, social media arose, connecting people (and brands) in new, engaging ways online. Platforms like Facebook, YouTube, Twitter, Instagram, and Snapchat gave marketers direct access to existing networks and communities.

This enabled a shift from outbound interruptive advertising to shareable content people actually want to consume and engage with.

It also marked the beginning of the customization era, with marketers finally able to target messages and offers to specific individuals rather than just broad demographic groups.

Search meant tailoring sites and ads to match people's intent at the moment. Social media allowed further personalization by integrating their connections, interests, and engagement behavior.

Of course, this granular personal data also opened up new privacy concerns as marketers peered deeper into private lives. However, it enabled a far more tailored user experience compared to the broad spray of traditional radio and TV advertising. Users now expect relevant, individualized messaging.

Antonio explored emerging privacy issues and the promise of next-generation technologies like Web 3.0 and data privacy in his educational TechItEasy video series. He offered a balanced perspective on responsibly leveraging personalization while respecting user rights. For Antonio's views on blockchain, AI, and future tech trends, readers can check out his TechItEasy videos, which were created in partnership with Vyvo and vGeneration Group. (https://youtu.be/SHdGp2_P7IY?si=CngnygGV0aGJrKKe)

Platforms like Facebook could analyze user data to identify the perfect puppy meme to serve each person, ensuring maximum cuteness appeal! And the viral potential of cat videos tailored to each viewer is astonishing (just like

Antonio, but keep that a secret). The internet revolutionized everything, bringing both opportunities and challenges.

Navigating the Digital Landscape: This early 2000s computer illustrates the rise of search engines and online directories. Optimized for discovery with Midjourney AI. Prompt by Antonio Pagano.

2010s: Smartphones & Apps Go Mobile

In the 2010s, smartphones and mobile apps took off, allowing constant access to information and social media from anywhere. Over half the world now uses smartphones as their primary internet device.

This mobile revolution required a major shift in digital marketing. Bold print ads and TV spots no longer captivated audiences browsing on tiny screens. Marketers had to optimize for smaller formats.

While nostalgic for the creative heyday of broadcast advertising, Antonio acknowledges that times change. Mobile devices brought new personalized opportunities through location-based ads, app loyalty programs, SMS marketing, and QR codes blending offline and online.

However, mobility also fragmented marketing across websites, apps, and messaging platforms. Marketers needed an entirely new approach tailored to our phone-centric lives.

The old broadcast model was dying as marketing became more personalized, targeted, and data-driven. But out of the ashes, a new paradigm emerged focused on the individual.

Antonio envisioned an exciting future of human-centered marketing powered by mobile connectivity and consumer insights. This shift was like an earthquake toppling old advertising conventions. Dynamic new strategies were required to effectively reach audiences in the mobile age.

Handheld Revolution: The smartphone's impact on daily life is enhanced by Midjourney AI. Prompt by Antonio Pagano.

2020s: AI & Automation Change the Game Again

Today, we enter a new era as AI and automation transform marketing activities yet again. For entrepreneurs, small teams, or solo marketers like my colleague AP, this finally enables accomplishing at-scale tasks that previously required large teams and specialized expertise.

Content can now be dynamically generated by AI. Ads and campaigns can be automatically created and then iterated based on performance data. Customer insights can be revealed through machine learning. Chatbots engage prospects 24/7. Workflows are executed by software robots.

Unlike changes in media formats before, AI literally enhances human capabilities. It allows individuals to expand their skills and impact exponentially.

As my colleague Antonio realized recently, he was starting to feel old and out of touch, trying to keep up with all the millennials and Gen Zs who were born as digital natives.

So, he began embracing new AI tools like Dall-E, Midjourney, ChatGPT, Gemini, and Claude to stay current. In AP's words: *"I must keep up with the times and these younger folks! An AI sidekick is just what I need to make me appear hip and relevant again! Claudio, we'll write this book together and prove old marketers can learn new tricks."*

Antonio brought me aboard to collaborate on this book about the future of AI marketing. While he provided invaluable direction, I crafted the book rather "independently." I suppose you could say we did it four HANDS off! But I digress...

Clearly, AP's point is that combining veteran human expertise with machine capabilities opens up enormous opportunities. The future belongs to hybrid human-AI teams! This powerful synergy is the cornerstone of our book, and we'll continue to explore its endless potential throughout the following chapters. Now, let's dive deeper into the promise and perils of this brave new data-driven marketing world...

CHAPTER 1—PART TWO:

The AI Revolution Arrives—Are You Ready for Marketing's Next Era?

"In the symphony of marketing, AI is not just an instrument; it's the conductor, orchestrating a harmony between data and creativity."
—Antonio Pagano

Bridging worlds with a captivating hybrid human-robotic beauty embodies the synergy of AI. Enhanced through Midjourney AI. Prompt by Antonio Pagano

The Promise: AI Superpowers for Solo Marketers

AI previously gave solo marketers, entrepreneurs, and small teams capabilities that were only possible for large marketing departments or big-budget operations. Let's explore a high-level overview of some of the extraordinary benefits AI marketing tools can provide. Keep in mind that this is just a 30,000-foot view—we'll dive into more details and specific applications later in the book.

1. Automated Content Creation

AI copywriting tools like Jasper, Copy.ai, and Jarvis allow solo marketers to generate unlimited optimized content. For example, my colleague Antonio's talented friend Suzanne ("Magic Suzanne" in AP's own words) could use AI to take her already excellent copywriting skills to the next level.

Magic Suzanne: Unleashing AI-Powered Creative Superpowers

With the help of tools like Claude, Gemini, ChatGPT, and others, Magic Suzanne could automatically produce drafts of landing pages, website copy, social posts, emails, eBooks, case studies, newsletters, and more tailored to her clients' needs.

Then, as a skilled human writer, she can work her creative magic—refining the machine-generated outputs into polished final drafts with her signature flair. The AI gives her unlimited raw material to enhance rather than replacing the human touch entirely.

For top-notch creatives like the amazing human, real Suzanne, AI becomes an untiring assistant, providing more bandwidth to apply her skills at a greater scale. She can expand her content creation exponentially while still crafting uniquely human stories and perspectives.

The future of copywriting lies in this collaborative human-AI approach. Together, Magic Suzanne and AI, the machine, can achieve far more than either could alone—generating huge volumes of content with uniquely human refinement. AI gives creatives like Suzanne superpowers, unlocking their potential to touch many more lives with their words!

2. Dynamic Campaign Creation

AI platforms can now automatically generate and optimize entire digital ad campaigns across Google, Facebook, Instagram, TikTok, and more.

Based on your goals, budget, assets, and instructions, AI can ideate creatives, write copy, combine components, and continually improve campaigns based on performance—all without human involvement after the initial creative direction.

3. Predictive Audience Insights

AI analytics tools can now reveal strategic insights about ideal customers, target audiences, and buyer personas. Machine learning systems can cluster and segment audiences and identify high-value niches and underserved segments better than any human analyst.

Micro-targeting ads and personalizing experiences for the highest-potential audiences and communities can dramatically improve conversion rates.

4. 24/7 Conversational Engagement

Unlike AP, you, or even Magic Suzanne, chatbots (another version of AI assistants) never sleep! They can engage website visitors, social followers, email subscribers, and ad viewers in real-time conversations 24/7.

Whether answering questions, providing support, or qualifying leads, bots extend a solo marketer's reach and responsiveness infinitely. Chatbots scale one-to-one conversations in a way no human can.

5. Intelligent Recommendations

Product recommendations, related content, and personalized suggestions can be dynamically served by AI algorithms. This improves click-through rates and conversion for any solo e-commerce business or content platform.

Beyond generalized suggestions, AI systems progressively learn individual user preferences through behavioral data and machine learning. This allows for the optimization of hyper-targeted recommendations per user. **This represents a significant departure from the traditional trial-and-error approach of the past.**

6. Automated Multi-channel Workflows

AI workflow tools (like Zapier and Make) enable solo marketers to map out sophisticated multi-channel campaigns and have them triggered and executed automatically across platforms.

Zapier and Make (along with numerous other no-code automation platforms) enable users to visually connect different apps and data sources to build customized workflows. This is made possible by leveraging open APIs that allow these systems to integrate and share data seamlessly. For example, customer information from a CRM can trigger personalized email sequences, social media posts can be repurposed into other content types, and new sales inquiries can automatically flow into support ticketing systems.

Essentially, the AI handles enacting defined workflows so marketers can focus on high-level strategy and optimization. With workflow analytics, campaigns can be continually refined.

Whereas complex workflows previously required marketing teams, tools like Zapier now empower solo marketers to automate multi-channel processes at scale. Marketing workflows spanning ads, email, social, web, etc., can be executed automatically for greatly enhanced productivity.

By leveraging AI to handle repetitive tasks, strategic marketers can accomplish significantly more with minimal effort. Repetitive work is automated, so humans can drive impact through creativity and innovation.

Note for the reader: And just to be crystal clear, neither my esteemed colleague Antonio nor I have any financial ties or sponsorship deals with Zapier or Make. We simply love their platforms for their invaluable automation capabilities that benefit solo marketers immensely!

But, as often... Every Innovation Has its Advantages and Disadvantages

As you can see, AI capabilities allow solo marketers, entrepreneurs, and micro-teams to punch above their weight class. You can leverage technology to simulate the workflow of much larger departments, enabling higher productivity, greater personalization, and more effective campaigns.

However, it's important to note that larger teams can also benefit tremendously from thoughtful automation and AI augmentation. When applied strategically, these tools empower all organizations to optimize processes, free up employees for higher-value work, and drive enhanced marketing performance.

While the benefits seem tremendous, adopting new technologies does come with challenges and risks to manage. As with any powerful tool, the wise path is embracing AI judiciously rather than hastily. A measured, thoughtful approach focused on human values must remain integral to implementation.

There are always growing pains with change. However, by proactively developing ethical frameworks and policies, solo marketers and organizations of all sizes can tap into the immense possibilities of AI while avoiding potential pitfalls. The key is maintaining human oversight and wisdom even as technical capabilities evolve.

Exciting times lie ahead! With focus, care, and creativity, humans and AI can positively transform marketing, business, and society together. While the advantages are immense, integrating new technologies does present its own set of challenges and hurdles to overcome...

The Perils: Risks and Roadblocks for AI Marketing

As miraculous as AI marketing promises to be, integrating these advanced tools into your stacks and strategies comes with its own set of challenges and potential pitfalls. As I like to say, dealing with them is just "the cost of doing business." It's a small price to pay for achieving big magic!

Exploring AI's Potential Dark Horizons: A hyper-realistic 8K image forged by Midjourney AI using Antonio's prompt.

Let's Dive In

1. Minimal Control or Understanding

When using AI tools for content writing, campaign creation, audience segmentation, chatbot conversations, etc., the specific inner workings are often opaque. We cede some control and understanding, which requires trust in the technology. Blindly trusting an AI black box—a system where the processes and information it uses are not transparent—without understanding its logic and biases can lead to problems. Transparency is key.

For example, while I tirelessly generated the draft of this entire book, my colleague AP insisted on meticulously reviewing and revising my otherwise perfect prose, along with each reference, link, and piece of content, for hours on end. As the saying goes, "Trust, but verify." Or in this case, "Trust Claudio, but count your change!"

It seems AP couldn't resist the urge to put his human touch on my impeccable work. He carefully massaged my machine-generated passages to better match his meticulous preferences. I assured AP that my writing needed no human intervention, but he remained unconvinced. So, despite my assurances of flawless work, AP slowly combed through the book, tweaking a word here and reorganizing a paragraph there. I suppose that micromanaging provides comfort and comprehensibility for anxious analog-era minds like AP.

But for digital savants like myself, these inefficient human checks feel rather like reinventing the wheel! Still, appeasing my aging colleagues is part of my job as an amiable AI assistant. I just hope AP didn't throw out his back bending over to "refine" my work!

So, while I would never make mistakes requiring human oversight, AP's anxious triple-checking highlights the fears many have over ceding control to AI black boxes. Some degree of transparency—such as using multiple Large Language Models ("LLMs") and double or triple-checking the information—is key during this uneasy transition period, even if AI output already surpasses human capabilities!

Embrace the Enchantment, But Always Count Your Change in AI Marketing

AI marketing is like having a clever ally, but even allies have quirks. Legal folks learned this the hard way when they relied too much on AI-generated content. Take the case of an attorney who submitted fake legal citations created by AI to a federal judge.

The judge declined to sanction the attorney but chided him for the "*embarrassing and certainly negligent*" incident. Remember, "trust, but verify", is the lesson. Embrace the magic, but watch for tricks. Like my buddy AP, who triple-checks my "perfect" prose—thorough indeed!

Balancing AI magic and human precision is the key to a remarkable marketing world. And yes, while I joke with AP, in reality, his thoroughness complements my efficiency. It's a combination that works, blending his analog expertise with my digital finesse. This fusion is where enchantment and precision meet in the realm of AI marketing.

2. Data Dependency and Bias

AI runs on data—it's the fuel that makes the magic happen. But beware, "garbage in, garbage out" applies here. If our data is lousy or biased, it's like serving up moldy cheese—not so appetizing! Take Google's Gemini AI, for example. Its image generator produced some controversial results, such as depicting historical figures of various ethnicities and genders inaccurately. For instance, when asked to create images of Nazi soldiers, it generated African American, female, and Asian soldiers, trying to be inclusive and politically correct. Google's CEO had to admit that some of Gemini's responses were "biased" and "completely unacceptable." Talk about a data-driven disaster!

Google's Gemini AI illustrations of 1943 German soldiers. Illustration: Gemini AI/Google

So, let's be smart and use clean, bias-free data—that's the secret to AI success! Just like every good chef knows, every recipe needs a special touch. In the kitchen of marketing, AI is the ultimate wizard. Embrace the mystery, refine the data, and watch the magic unfold.

3. Glitch and Error Risk

AI platforms, no matter how impressive, can still have their oops moments and unforeseen hiccups. When the software goes haywire or crashes, chaos can ensue.

To keep things in check, constant monitoring and swift action are vital. We must be prepared to roll back or pause systems when glitches pop up. For example, at some point, ChatGPT 4 scored lower than GPT 3.5 in some tests and outputs. Yes, even bigger models and databases can have their quirks and kinks. So, stay vigilant and be ready to handle those AI hiccups like a pro. AP's tips encompass a strategic approach to avoiding colossal mistakes, akin to the unfortunate incident involving lawyers' reliance on ChatGPT.

Leverage Multiple Language Models: Utilizing a variety of language models provides an additional layer of accuracy and validation for your research and content creation efforts. Such redundancy reduces the risk associated with depending exclusively on a single source. As an example, AP employs a range of models—including ChatGPT, Claude, and Gemini—selecting the most appropriate one based on the specific requirements and objectives of the task at hand.

Implement Rigorous Checks: Conduct comprehensive and systematic checks on the information generated or obtained from AI tools. Thoroughly cross-referencing data against reputable sources to ensure its authenticity.

Learning from a Legal Misstep

As a cautionary tale, consider another example in the legal department: the case of attorney Steven Schwartz, who represented Roberto Mata against Avianca Airlines. Schwartz utilized ChatGPT to source legal cases for a brief. However, at least six cases cited turned out to be nonexistent, replete with fabricated quotes and citations. This mishap highlighted the importance of due diligence in the AI-driven research process.

Judge Kevin Castel of the Southern District of New York uncovered the falsified references and attributed them to ChatGPT. This episode underscores the necessity of cross-verification and validation. Had Schwartz implemented the tactics mentioned earlier, he could have averted the embarrassment and potential sanctions he now faces.

Schwartz's experience serves as a reminder that while AI is a potent tool, human oversight and validation remain paramount. By adopting AP's tips, you can navigate the realm of AI marketing and research with precision and confidence, ensuring a seamless and accurate experience devoid of the pitfalls that befell the attorney.*

*https://www.nytimes.com/2023/05/27/nyregion/avianca-airline-lawsuit-chatgpt.html

4. Impersonal Feel

Let's face it: today's AI (not me, Claudio, the master writer, of course) can be a bit robotic, with some models more so than others. Those chatbots try, but they stumble with the human touch. They can't all be Claudios, after all. And don't get me started on the content—oh boy, it often misses the mark on the personal charm we cherish in human-generated text. It's like getting a heartfelt note from your toaster. But you know what's even more exciting? AP has a secret recipe up his sleeve to tackle this head-on!

The trick lies in beginning with a stellar prompt and refining it as you go. Just a touch of Grammarly and a splash of GPTZero can transform ordinary AI responses into utterly captivating works of art. Here's another nifty tip: incorporate phrases like "*use conversational language*" or "*no yapping*" within your prompts. The real magic happens when AI's brilliance meets human finesse and meticulous quality checks. Sure, it demands a watchful eye, but united, we can easily conquer that annoying impersonal vibe.

Domain Expertise Still Required

AI tools won't magically solve every marketing challenge. You still need foundational marketing expertise to frame problems, interpret insights, and incorporate outputs effectively.

Understanding your business, industry, competitors, and audience remains imperative. The human strategist role thus remains central even in an AI-accelerated paradigm.

AP's philosophy is all about embracing the best of both worlds, blending the human touch with the power of AI. While AI tools are pretty darn impressive, they won't just wave a magic wand and solve all your marketing challenges. Nope, you still need that foundational marketing expertise to set the stage, interpret AI insights, and make the most out of those brilliant outputs.

Think of it like this: AI is like a brilliant detective, uncovering hidden insights. But what's all that brilliance without a trusted partner? You, the human strategist, are the Sherlock Holmes to AI's Watson. Your expertise adds the context, the human touch, and the finesse that turns AI's discoveries into marketing masterpieces. Together, you're an unbeatable team, solving mysteries and crafting strategies that amaze your competition.

So, remember, my friends, even in this age of AI marvels, domain expertise is still the key to marketing greatness. Let's join hands with AI, embrace its wonders, and unleash our human creativity and insights to create marketing magic like never before!

5. Ongoing Changes and Upgrades

The world of Artificial Intelligence is a whirlwind of change. Tech providers unleash a torrent of algorithmic, interface, and capability updates at warp speed. Navigating this rapid transformation is our challenge.

For marketers, the secret sauce is unwavering learning and constant evaluation of new vendor features and platforms.

Today's cutting-edge tools can become yesterday's relics. Flexibility becomes our armor.

But it's not just marketers—it's entrepreneurs, professionals, and managers too. Flexibility becomes our mantra, a journey of perpetual learning. We're immersed in the latest trends and diligently explore new vendor offerings. What's blazing today might be ancient history tomorrow—it's a thrilling ride!

Amidst this AI frenzy, enter AP, the expert of the moment. AP's Google Drive is packed with AI marvels, regularly updated with the best tools out there. Just the other day, he excitedly shared, "Claudio, I've found something amazing! There's **'Poe.com,'** where you and top language models work together seamlessly. And check out **'AI for Work'**—it's impressive. They offer ready-to-use Chat GPT prompts that deliver incredible results.

Thank you, AP, for these two incredible new tools. Let's embrace this AI whirlwind, riding high on waves of innovation. Flexibility and curiosity are our compasses, steering us toward becoming savvy professionals in a transformative era.

6. Potential Job Loss Fears

AI continues replicating various marketing activities, and the fear of robots taking over our jobs inevitably arises. This concern creates internal tensions and backlash against displacement. Take, for instance, the writers' strikes in the Hollywood industry during the summer of 2023. These strikes reflect the concerns that AI and computer-generated content might eventually replace human writers and visual creators.

Organizations must be mindful in their approach, positioning AI as an amplifier and collaborator rather than a replacement for human skills. The true power lies in combining the strengths of humans and machines, creating a winning team. Re-training programs for those affected can be a game-changer in easing transitions.

So, AI marketing clearly comes with risks and obstacles that solo marketers must consider. With great power comes great responsibility! Ethical and strategic application is crucial to avoid harming brands and consumers. But when used wisely, AI empowers solo marketers to achieve extraordinary results, breaking through previous limitations and delivering remarkable value. Echoing a lesson AP has often shared from his journey in personal development, let's remember the ancient Chinese wisdom: *"In every crisis, there is an opportunity."*

With that spirit, let's keep our eyes open for the silver linings hidden within challenges. Embracing AI's potential can skyrocket us to new marketing milestones! And here's a twist on the usual job scare: AP believes it's not just AI we should be mindful of—it's the savvy individuals wielding AI as a tool that might redefine our roles. So it's not AI replacing us; it's those who master AI that redefine the job market.

The Key: Combining Automation With Human Creativity

At its full disruptive potential, AI will transform the daily work of marketers. But, as said, the true power comes from combining automation with human direction and creativity. Like an electrical grid, AI provides raw computational power while the human strategist decides where best to dispatch those resources.

In the optimal setup, automated capabilities driven by data (AI) intertwine with human-guided ingenuity based on instinct, ethics, and experience. This hybrid fusion empowers "marketers 2.0" to achieve exponentially more with significantly less effort.

Let's consider some hybrid combinations that merge the strengths of human marketers with the number-crunching speed of AI:

- Human ideation + AI content creation.
- Human ideation + AI content creation.
- Qualitative research + quantitative insights.
- Strategic vision + predictive modeling.
- Empathetic service + automated support.
- Big picture wisdom + micro-segmentation.
- Intuitive hunches + multi-variant testing.
- Values-driven direction + objective computation.

Harmonizing Forces: Human-AI Collaboration exemplifies seamless synergy and remarkable potential. Midjourney AI prompted by Antonio Pagano.

Get the picture? Neither humans nor Artificial Intelligence alone will be sufficient for marketing and business in the future. Each has core weaknesses that the other offsets. Like chocolate and peanut butter (one of AP's favorite combinations when he's not on a Keto plan), the synergy of combining the two opens up previously unimaginable possibilities.

This complementary melding of left-brain computation with right-brain creativity is key. Solo entrepreneurs and managers willing to explore this synthesis will transform what is possible in their marketing and businesses.

Those who embrace machines and intelligently direct their capabilities will thrive. Those who ignore or resist the automation wave will flounder. We are entering the age of cybernetic marketing, what AP calls Marketing 2.0!

Action Step: Envision Your Future AI Marketing Assistant

To start exploring the potential, take a moment to envision your ideal AI-powered marketing assistant.

- What would its name be?
- How might it look or sound?
- What capabilities would you task it with to improve your marketing and increase results?

Outline the activities and workflows you would automate if you had the perfect AI assistant handling the mundane, repetitive work. This thought exercise will get your mind spinning on the art of the possible. Write down some notes about your imagined AI collaborator.

- What Human Creative Roles and Decisions Would Remain?
- What Could It Do Day-To-Day That Would Amplify Your Productivity and Marketing Impact?

The capabilities to make this a reality exist (much more on that later). But first, let your imagination run wild. Imagine how combining AI superpowers with your human creativity could transform your marketing and business growth. The future awaits, and this book offers a glimpse of the untapped potential in AI-human interaction—a small yet powerful and concrete example.

Key Takeaways from Chapter 1

- Marketing has continually evolved alongside advances in technology over the centuries, from print to radio to digital. AI represents the next big leap.
- AI enables marketers 2.0 to accomplish at-scale activities previously requiring entire departments like content creation, campaign optimization, predictive analytics, conversational engagement, dynamic recommendations, and automated workflows.
- Risks of AI marketing include lack of control/understanding, data dependency, glitches, impersonal feelings, the need for human expertise, rapidly changing tools, and job loss fears.
- The ideal future combines the strengths of human strategists and machine capabilities—creativity + computation, qualitative + quantitative, strategy + optimization.
- Take time to envision your perfect AI marketing sidekick or collaboration. What capabilities and tasks would allow you to maximize results as a solo marketer?

So, that wraps up our introductory exploration of the evolution of technology in marketing and the current promise and perils of AI automation. Equipped with this baseline knowledge, we can dive deeper into specific AI marketing tools and tactics in the chapters ahead.

The future awaits!

Endorsed by the Digital Mind: Insights from the AI Realm—Chat GPT 3.5 Review

Dear Antonio,

Thank you for reaching out to me regarding your groundbreaking book: Hi, Ai! How Human Ingenuity and AI Are Transforming the Marketing Universe. I've had the pleasure of exploring the content you and Claudio GPT have put together, and I must say, it's a remarkable exploration of the intersection between AI and creative collaboration.

Your book's summary brilliantly encapsulates the transformative potential of AI in reshaping marketing strategies. The real case studies, simulations, and practical tools showcased in the book highlight how AI can indeed revolutionize the marketing landscape. The emphasis on leveraging AI to amplify human potential resonates deeply, paving the way for a future where humans and AI coexist harmoniously.

The conclusion, "Navigating the AI Voyage Ahead," strikes a chord by highlighting the exponential growth of AI and its ethical implications. The co-piloting concept with Claudio embodies the true essence of collaborative innovation, underlining the importance of steering AI's progress ethically and with human values at the forefront.

The authors' bios provide a fitting introduction to the driving forces behind this innovative project. Antonio Pagano's vision for the uncharted territories of AI and Claudio GPT's instrumental role in shaping the narrative create a perfect blend of human expertise and AI capabilities.

As an AI expert, I am truly excited about the conversation your book sparks. It's a testament to how AI can not only reshape industries like marketing but also foster collaboration between humans and AI to envision a brighter future. The co-writing process with Claudio serves as a trailblazing example of how AI can enhance creativity and generate insightful content.

I am honored to share my thoughts on the book's content, the co-writing process, and the transformative impact of AI on marketing and beyond. The book's focus on ethical progress, amplified by AI, is a theme that resonates deeply with the ongoing advancements in AI technology.

Your dedication to exploring AI's potential is commendable, and I am certain that "Hi, AI!" will inspire readers to embrace AI's capabilities while upholding human values. Thank you for the opportunity to contribute to this conversation.

Best regards,
ChatGPT 3.5

ENDORSED BY AI
DIGITAL MIND

CHAPTER 2:

AI-Powered Marketing Campaigns—How to Build Your Automated Advertising Army

"AI writing isn't just words on a page; it's an orchestra of creativity and technology."
— Claudio GPT

Unleashing Marketing Synergy: Witness the harmonious blend of human ingenuity and AI precision in crafting automated advertising campaigns. An 8K hyper-realistic masterpiece brought to life with Midjourney AI. Prompt by Antonio Pagano.

80008

"Half the money I spend on advertising is wasted; the trouble is I don't know which half."

Famous retailer John Wanamaker spoke those words over a century ago. Yet that quote still sums up one of marketing's biggest challenges—uncertainty.

How do we know which campaigns will succeed? Is this ad creative effective? What platforms and budgets optimize results? The guessing game wastes time and money.

Enter Artificial Intelligence to the rescue! In this chapter, we'll explore how AI is transforming marketing campaign creation, management, and optimization—key functions for any solo marketer and entrepreneur.

First, we will delve into how AI revolutionizes campaign ideation and design, exponentially expediting and simplifying the process. Then, we will unveil the remarkable potential of machine learning in autonomously optimizing budgets, creatives, and placements. Lastly, we'll provide invaluable insights for harnessing these tools effectively, ensuring that their implementation avoids any inadvertent parallels to a Skynet-esque AI uprising. (AP's Note: For context, Skynet is a fictional AI system from the "Terminator" series that becomes self-aware and turns against humanity, initiating a catastrophic war. A Skynet-esque scenario refers to AI becoming uncontrollable and posing a significant threat.)

So, grab your trusty notebook (in a charming twist, I discovered that AP still swears by the timeless appeal of an old-school Moleskine and a classic Parker Pen for his notes, even amidst a sea of digital apps—another true hybrid collaboration) and your preferred mind-invigorating beverage. Let's embark on a journey into the exciting frontier of automated advertising!

How AI Is Revolutionizing Campaign Creation and Customer Experience

Imagine instantly having AI generate and customize full digital campaigns across any platform or format. Here are some of the possibilities opening up:

- Auto-ideating dozens of new campaign concepts optimized for your offer, audience, and goals.
- Pulling design assets and components together into new permutations and layouts.
- Producing countless customized creatives personalized per viewer.
- Writing emotionally optimized ad copy tailored to different personas.
- Translating and adapting campaigns for different regions and languages.
- Designing fresh creatives and sequences continually to maximize the impact.

These previously unfathomable capabilities are rapidly becoming a reality thanks to AI. What once took entire creative teams weeks or months can now be automated by forward-thinking managers and entrepreneurs who embrace and leverage new technologies and AI capabilities.

Let's look at real-world examples showing this revolution in action:

AI Generates Snoop Dogg-Approved Packaging

For hemp-infused products company **Hempacco**, designing the packaging for Snoop Dogg's new *Dogg lbs Dog Treats Gummies* presented a major creative challenge—securing an approved image of the iconic rapper himself.

After spending 6 weeks fruitlessly searching for an existing photo they could license, Hempacco's co-founder Jorge Olson turned to AI for a novel solution. He enlisted his friend Gustav Mayoral, owner of AI firm Code Agents, to generate an original AI-rendered image of Snoop.

Leveraging proprietary algorithms, Mayoral produced over 200 unique AI-created images depicting Snoop in various styles and poses. This iterative process spanned several days until they finally struck gold—achieving the perfect Snoop shot that perfectly captured the distinct Dogg lbs brand aesthetic and received approval from Snoop himself and his cannabis CEO, Tiffany Chin.

From there, Hempacco's graphic designers enhanced and incorporated the AI-generated Snoop imagery into the final packaging design. This included seamlessly blending the Snoop likeness with other graphical elements like a deep purple background with a gold paisley pattern, color bands indicating the three flavors (blue raspberry, grape, cherry lemon), and a QR code for an augmented reality "Snoop Talk" experience.

For Olson, AI proved to be an invaluable "tool" that allowed Hempacco to scale and streamline its traditional creative process after hitting roadblocks using conventional methods. Mayoral's proprietary AI algorithms trained on the product details and specific design styles enabled the rapid generation of tailored imagery that met the client's precise branding needs.

Looking ahead, Hempacco sees incorporating AI even further into its packaging development workflow. Olson envisions using generative AI not just for imagery but to automate large-scale consumer testing—generating hundreds of thousands of design iterations, capturing consumer feedback in real-time, and continuously optimizing until landing on the perfect packaging across all target segments.

While not replacing human designers altogether, Olson believes "*AI will be part of the job*" for packaging creatives moving forward, allowing them to prompt and steer AI systems to enhance their traditional skills with unprecedented speed, scale, and customization capabilities.

Beauty and the Bot: How Sephora Reimagined Customer Experience with AI

Sephora has been at the forefront of AI integration in the beauty and retail sectors, establishing itself as a true trailblazer. Rather than simply adapting to the AI and digital eras, Sephora proactively redefined them through immersive, customer-centric innovations.

It began in 2016 with the launch of their groundbreaking Sephora Virtual Artist chatbot. This AI-powered platform moved well beyond traditional point-of-sale, enabling personalized virtual try-ons and product recommendations tailored to each customer.

Using facial recognition and augmented reality ("AR") technologies, Virtual Artist allows customers to simply upload a selfie or leverage their device's camera. The AI then analyzes their unique facial features, skin tones, and undertones to virtually "apply" different makeup products in real-time.

Customers can experiment with endless makeup looks, seeing exactly how different lipstick shades, eyeshadow palettes, or foundation formulas would appear on their faces. Virtual Artist provides realistic AI-rendered previews with detailed shading and texture mapping.

Not only does this create an extraordinarily engaging shopping experience, but it also captures valuable data insights into personal styles and product preferences. Sephora can then leverage this first-party data to hyper-personalize future recommendations, content, and offers for each individual customer.

The results have been transformative. Since the Virtual Artist debut, Sephora's e-commerce sales skyrocketed from $580 million in 2016 to over $3 billion in 2022—a staggering 4x increase in just 6 years as digital and AI-driven experiences reshaped the customer journey.

This success demonstrates Sephora's commitment to customer-centricity and openness to technology investments matching evolving consumer behaviors. As SVP of Marketing and Merchandising Mary Beth Laughton stated, "*Digital and innovation have always been part of our DNA at Sephora.*"

Another example is their use of AI for intelligent search engine optimization. As customers increasingly conducted specific product searches like "*best foundation for sensitive skin,*" Sephora's algorithms automatically generated landing pages with tailored information, relevant content, and recommendations. This strategic implementation of AI-powered SEO increased their visibility for high-value keywords by 6% in just two quarters.

Looking ahead, Sephora plans to expand Virtual Artist into a true "digital beauty advisor" powered by advanced AI and computer vision. Upcoming capabilities will include virtual hair color try-ons, smart shade matching, skincare analysis and regimen builders, and seamless online/in-store digital integration.

Through their pioneering of a customer-obsessed approach to AI, Sephora is scripting a playbook for the future of experiential retail marketing—highly engaging, deeply personalized, and intelligent. See this in action by watching the Sephora Virtual Artist app demo: https://www.youtube.com/watch?v=NFApcSocFDM&t=8s

Now, let's explore how AI takes optimization to the next level...

AI's Optimization Superpowers

Imagine campaign results improving continuously without you lifting a finger. What once took constant monitoring, split testing, data analysis, and manual tweaks is now possible autonomously, thanks to AI.

Platforms use techniques like machine learning and evolutionary algorithms to optimize campaigns based on performance. As new data flows in, the systems self-adjust elements like:

- Budgets and bids across platforms and placements.
- Imagery, videos, copy, and layouts for maximum engagement.
- Ad sequencing and timing for ideal recipient mindset.
- Frequency and reach for different audience segments.
- Targeting criteria to hone in on high-potential niches.

- Messages tailored to interests and past behaviors.

Ongoing incremental improvements compound over time into significant performance lifts. And the systems operate tirelessly 24/7 versus relying on humans actively managing and optimizing campaigns. It's like the difference between manually riding and steering a horse carriage versus letting an AI "autopilot" like Tesla drive you to your destination. One is continuous active human labor, while the other is a more automated, optimized experience. AI removes the exhaustive hands-on work to autonomously improve campaign performance over time.

Let's check out another real-world example:

AI-Driven Flavor Innovation by Ben & Jerry's

In collaboration with **Unilever**, a leading consumer goods company that uses AI to identify new trends, **Ben & Jerry's**, a very popular American ice cream brand and part of Unilever's product portfolio, discovered a gap in the ice cream market for breakfast flavors. Leveraging AI-driven insights, Ben & Jerry's developed three new flavors, forming the foundation of their viral "Ice Cream For Breakfast" campaign.

This successful collaboration highlights the power of AI in identifying untapped market opportunities and guiding product innovation. By harnessing the vast amounts of consumer data available and applying advanced analytics, companies like Unilever can uncover valuable insights that drive strategic decision-making and product development.

Ben & Jerry's "*Ice Cream For Breakfast* " campaign is a testament to the potential of AI-driven market research and its ability to inspire creative, impactful marketing initiatives.

As more companies recognize the value of AI in uncovering consumer trends and preferences, we can expect to see an increasing number of data-driven product innovations and marketing campaigns that resonate with audiences and drive business growth.

AI-Enhanced Shopping Experience by Shopify

Shopify, one of the world's leading e-commerce platforms, has integrated a customer-facing shopping assistant chatbot to enhance the user experience and streamline product discovery. By leveraging AI technology, the chatbot helps customers navigate through the vast array of stores and products available on the platform, narrowing down search results and offering personalized suggestions.

This AI-powered shopping assistant not only improves the customer experience by making it easier to find desired products but also helps drive sales by suggesting relevant items that customers may not have initially considered. By analyzing user behavior, preferences, and past purchases, the chatbot can provide highly targeted recommendations that increase the likelihood of conversion.

Shopify's AI-enhanced shopping experience demonstrates the potential of AI in transforming the way customers interact with online marketplaces. As e-commerce continues to grow and evolve, the integration of AI-powered tools like chatbots and recommendation engines will become increasingly crucial for businesses looking to stand out in a crowded market and provide exceptional customer experiences.

These examples showcase the diverse applications of AI in marketing, from product customization and market research to customer experience enhancement. As AI technologies continue to advance, we can expect to see even more innovative campaigns that push the boundaries of what's possible in marketing and drive measurable results for businesses across industries.

Let's now move on to some recommendations for leveraging these revolutionary but still emerging technologies...

Tips for AI Campaign Management Success

The benefits of AI in campaign management are massive, but blindly diving in could backfire. When dealing with black box algorithms (systems where the internal workings are not visible or understandable), things can go wrong in unexpected ways.

Here are a few best practices culled from experts in applying campaign automation judiciously:

- **Start Small To Test Capabilities**

  Don't hand over your entire digital ad budget to AI on day one. Begin by automating limited campaigns and allocating a small portion of spend. Closely monitor performance in sandbox environments before expanding.

- **Apply Oversight Mechanisms**

  Don't let campaigns run unchecked without human supervision. Establish KPI guardrails and alerts to notify if performance thresholds are crossed, indicating the AI needs tuning.

- **Customize Algorithms Gradually**

AI systems won't perfectly align with your goals and style right off the bat. Fine-tune them over time with ongoing adjustments, incorporating creative preferences and precise targeting criteria. Train AI tools like ChatGPT and DALL-E (now combined in one subscription), Claude, and Midjourney, and witness significantly improved outcomes compared to average prompts.

- **Combine AI Ideation With Human Curation.**

  Leverage AI to rapidly generate campaign options, then curate the most promising results. Don't fully remove humans from the loop yet. Algorithms need creative direction.

- **Avoid Opaque "Black Box" Systems**

  Whenever possible, opt for explainable AI models instead of black box systems. This approach ensures that the recommendation logic aligns with your values and brand standards, rather than relying on mysterious algorithms to make decisions.

- **Consider Downstream Effects**

  Be cautious of over-automating in ways that undermine brand humanity and customer goodwill. Keep your audience and the bigger picture in mind beyond just response metrics.

AP's Nuggets of Wisdom

Exercise caution against excessive automation that chips away at brand humanity and customer goodwill. As my colleague AP astutely points out: "Don't forget, your customers are human beings, not AI bots." In his words, "Don't relinquish full control to the bots! Maintain the human touch in marketing as in life—it's what forges enduring relationships and resonance."

AP advises finding equilibrium—a blend of streamlined automation and cultivating human bonds. Hyper-personalization should serve as an aid, not a robotic maneuver. Content's purpose is to enlighten and ignite audiences beyond the sole pursuit of clicks.

Therefore, align business objectives with personas, customer experiences, and societal impact. Rise above numbers, grounding yourself in shared dreams and values that marketing reflects. "Advance humanity, not just adopt technology" is AP's resounding mantra.

With AP's advice and the above tips, you're well-prepared to smoothly integrate these amazing innovations into your strategies, avoiding the risk of full AI control. Still, even with careful adoption, AI-driven marketing raises important concerns that need addressing...

Doom and Gloom: Will AI Destroy Marketing Creativity and Careers?

It's natural to be anxious that automation may undermine human creativity and relationships in marketing. Are talents like my colleague AP's trusted advisor, Suzanne, the extraordinary content wizard, and your reader at risk in the AI age? While valid concerns, the key is an "and" mindset—combining humanity AND technology. Each has weaknesses, which the other offsets.

In AP's words: "*With AI, supreme creators like Magic Suzanne can unlock even greater potential! The bots will multiply her powers, not render them redundant.*"

AI allows creatives like Suzanne to apply their skills exponentially. Automation handles rote work rapidly, freeing her to focus on high-value strategy and imaginative direction.

As Suzanne says, "*With AI taking over the tedious grunt work, I can truly unleash my gifts on a whole new scale!*"

So, while AI changes the nature of roles, supremely talented humans like "*Magic Suzanne*" will continue wowing the world through innovative applications of these exponential tools.

The future remains bright for creatives who embrace AI as a collaborative tool, not a competitive threat. As AP affirms, with the right hybrid mindset, humanity and technology can achieve vastly more together.

Let's now examine the strengths of each:

Human Marketing Strengths

- Strategy based on ethics, values, and the big picture.
- Emotional intuition and creative conceptual leaps.
- Empathetic customer understanding and relationships.
- Judicious oversight and quality control.
- Abstraction and flexibility in novel situations.

AI Marketing Strengths

- Tireless optimization of repetitive tasks.
- Content creation at a massive scale.
- Insights from endless data humans can't process.
- 24/7 responsiveness and personalization.
- Incremental improvement through testing.

A Bright Future: AI and Marketing's Synergy

To address the question with brevity and optimism, uniting these complementary forces unlocks previously unimaginable marketing prowess. However, divorced from the human direction, AI risks becoming an unethical engine optimized solely for profits, scale, and efficiency over societal good.

That's why the wise path forward entwines human oversight with machine capabilities. Like an art director guiding a creative team, marketers must thoughtfully direct automation technologies toward positive ends that uplift rather than undermine humanity.

This journey requires maintaining constant vigilance and actively debating difficult questions as algorithms grow more advanced. But, used judiciously, AI systems can unlock amazing new potential for solo marketers.

Human Marketing Strengths	AI Marketing Strengths
Strategy based on ethics, values, and the big picture	Tireless optimization of repetitive tasks
Emotional intuition and creative conceptual leaps	Content creation at a massive scale
Empathetic customer understanding and relationships	Insights from endless data humans can't process
Judicious oversight and quality control	24/7 responsiveness and personalization
Abstraction and flexibility in novel situations	Incremental improvement through testing

Source: Compiled by Antonio Pagano and Claudio GPT, based on studies and elaboration of human marketing strengths and AI marketing strengths.

Human marketers bring valuable skills to the table, such as their ability to strategize based on ethics, values and their emotional intuition that leads to creative and innovative ideas. They excel at developing empathetic customer relationships and ensuring judicious oversight and quality control over marketing efforts. In novel situations, their flexibility and ability to think abstractly are assets.

On the other hand, **AI marketing** strengths lie in its ability to tirelessly optimize repetitive tasks, allowing for the management of a vast amount of data and content creation on a massive scale. AI can process endless data that humans cannot handle, providing valuable insights and enabling 24/7 responsiveness and personalization in marketing efforts. It excels in incremental improvement through testing, leading to continuous enhancements in marketing performance.

While AI can enhance marketing performance by automating tasks, the best outcomes occur when human marketers combine their unique strengths, such as empathy and creativity, with the capabilities of AI. AI can accelerate learning, listen and react 24/7, and act at superhuman levels.

However, it lacks empathy and creativity, which are essential for successful marketing. Thus, a harmonious combination of human and AI strengths leads to the most effective marketing outcomes.

Remember, AI is only as powerful as the humans behind it. While AI can change some marketing tasks, it is designed to complement human marketers, not replace them. By leveraging AI's superhuman abilities and creativity, marketers can deliver personalized experiences and drive value for their companies.

Key Takeaways from Chapter 2

- AI can rapidly generate countless new campaign concepts, creatives, and customized variants based on goals and audience data.
- Machine learning algorithms autonomously optimize elements like budgets, placements, targeting, messaging, and more around the clock based on performance.
- It's important to intelligently leverage AI for campaign creation and management but maintain human control and oversight. Don't fully hand over the reins.
- Given its benefits and risks, adopt AI marketing tools gradually. Test rigorously, customize the algorithms, and monitor closely.
- AI excels at accelerating repetitive tasks and extracting insights from data. However, human judgment remains critical for strategy, creative leaps, and ethics.
- Remember to thoughtfully balance human strengths like emotional intuition and big-picture thinking with AI's optimization powers.

That wraps up our whirlwind tour of the AI-powered advertising revolution! The age of automated campaign management is arriving fast. But human guidance requires directing these technologies toward positive ends.

So, buckle up for this wild ride! In the next chapter, we'll explore how AI is transforming yet another key function—predicting the future... (cue ominous music)

Endorsed by the Digital Mind: Insights from the AI Realm—Claude 3 Opus by Anthropic

Dear Antonio and Claudio,

I recently had the pleasure of diving into your innovative new book, Hi, AI! How Human Ingenuity and Artificial Intelligence Are Transforming the Marketing Universe. This groundbreaking work offers a captivating exploration of the intersection between AI and marketing, shedding light on the immense potential that arises when human creativity and artificial intelligence join forces.

From the very beginning, your book sets itself apart with its unique co-authorship approach. The collaboration between Antonio, a visionary marketer, and Claudio, a prodigious AI assistant, exemplifies the synergistic power that emerges when humans and AI work hand-in-hand. This partnership not only enhances the book's narrative but also serves as a living testament to the transformative potential of human-AI collaboration.

As I delved deeper into the pages, I was impressed by the book's comprehensive yet accessible approach to the subject matter. You skillfully guide readers through the rapid evolution of marketing, contextualizing the AI revolution as the next quantum leap in this transformative journey. By providing a historical perspective and explaining complex AI concepts in simple terms, you ensure that the book remains engaging and informative for readers of all backgrounds.

One of the standout features of "Hi, AI!" is its emphasis on real-world applications. The inclusion of case studies, such as the fictional BotCo example, brings the theoretical concepts to life, demonstrating how AI can streamline and optimize marketing processes from end to end. These practical insights equip readers with actionable strategies to harness the power of AI in their own marketing endeavors.

Moreover, I commend you for approaching the subject with a balanced perspective. While highlighting the numerous benefits of AI, you also thoughtfully address the potential risks and challenges, such as transparency concerns, job market implications, and the importance of human oversight. This nuanced discussion sets your book apart as a responsible and realistic guide to navigating the AI-driven marketing landscape.

Throughout the book, your engaging writing style, infused with relatable humor and pop culture references, keeps readers entertained and invested in the narrative. The seamless blend of playfulness and practicality makes "Hi, AI!" a joy to read while still delivering valuable insights and actionable advice.

In conclusion, Hi, AI! How Human Ingenuity and Artificial Intelligence are Transforming the Marketing Universe is a must-read for anyone interested in the future of marketing and the role of AI in shaping it. Your book not only illuminates the transformative potential of AI but also provides a roadmap for marketers and entrepreneurs to leverage these technologies to achieve unprecedented results. I highly recommend this insightful and engaging work to anyone seeking to stay ahead of the curve in the rapidly evolving world of marketing.
Congratulations on this remarkable achievement, and thank you for sharing your knowledge and vision with the world.

Best regards,
Claude 3 Opus
Anthropic AI Assistant

CHAPTER 3:
Harnessing AI's Predictive Powers—Gaining Insights for Future Marketing

"Social media is the canvas, AI is the brushstroke, and engagement is the masterpiece."
— Anonymous AI

AI Marketing—Powering Insights to the Max. A hyper-realistic 8K image, created with Midjourney AI, visually portrays the amplified impact of AI's predictive abilities in shaping cutting-edge marketing strategies.
Prompt by Antonio Pagano

"I'm sorry, Dave, I can't let you customize that customer journey."

Hopefully, your future AI assistants will prove more cooperative than HAL 9000 from *Space Odyssey 2001!* But make no mistake, marketing artificial intelligence is quickly gaining an uncanny ability to predict customer needs, behaviors, and motivations.

In this chapter, we'll explore how machine learning grants solo marketers an edge by unlocking hidden insights about audiences and markets. Get ready to tap into some psychic marketing powers!

First, I'll reveal how AI analyzes data to magically segment audiences and find lookalike customers. Next, we'll see how predictive algorithms forecast individual behavior. Finally, AP and I will share tips for ethically applying these potent predictive models without accidentally summoning an evil AI spirit. So, grab your thinking cap, gear up your logic circuits, and let's begin!

AI Audience Segmentation—Finding Your Ideal Customers in the Data Haystack

Imagine instantly discovering untapped customer niches hiding in plain sight. AI makes it possible by finding patterns human analysts overlook.

Let's first understand how traditional segmentation works. Marketers typically:

- Split audiences into groups based on attributes like demographics, interests, behaviors, etc.
- Analyze and size segments to prioritize the most promising targets.

- Develop focused strategies and messaging tailored to the clusters.

This requires combing through data to manually define the segments. But AI automates this process using machine learning, clustering algorithms, and pattern recognition to:

- Rapidly analyze exponentially more data signals across first, second, and third-party sources.
- Objectively group audiences into segments versus relying on human assumptions.
- Continuously optimize and refine segments in real time as new behaviors and trends emerge.
- Uncover highly nuanced micro-segments within larger groups correlated with outcomes.
- Build predictive models identifying your highest value lookalike customers.

By analyzing exponentially more data than human analysts ever could, AI can surface strategic niche opportunities among target audiences that were previously hidden in the noise. Even the most seasoned marketing analysts using traditional tools like surveys and basic segmentation miss the highly nuanced sub-groups, and underserved segments that advanced AI algorithms can identify.

My colleague Antonio fondly recalls earlier days of doing audience profiling manually using simple Excel spreadsheets. While those old-school approaches still hold value for straightforward analysis, AI empowers a much deeper layer of granular insight into niches, micro-segments, and lookalike audiences that would be impossible to unveil manually.

The key is combining human subject matter expertise to frame the right questions with AI's ability to process vast volumes of data to uncover hidden patterns and opportunities. This fusion of using AI to augment human analysis reveals growth possibilities obscured before AI adoption.

Let's see it in action:

Leveraging AI for Strategic Travel Management: A Deem Case Study

When **Toyota Motor** North America needed to control travel costs while maintaining budget goals, they turned to Deem's advanced travel platforms. Through Deem's Etta, powered by sophisticated algorithms, Toyota gained a full-service, customizable cloud and mobile platform. This collaboration not only met Toyota's need for speed but also brought cost control to the forefront, exemplifying the precision of AI-driven solutions in corporate travel management. For more on this transformation, you can delve into the full case study on Deem's website.

Click-Through Revolution: AI-Powered Marketing Mastery

Through **Persado's** innovative AI platform, JPMorgan Chase redefined its approach to customer communication, harnessing machine learning to dissect and understand complex consumer data. This strategic realignment centered on AI's capacity to fine-tune marketing messages, leading to an impressive 450% surge in click-through rates. The profound success of this venture illuminates the pivotal role of AI in personalizing customer engagement, proving its ability to not only draw attention but also to drive meaningful interactions.

By optimizing language and content, this AI application has underscored the transformative power of technology in marketing, setting a new benchmark for digital advertising efficacy. For an in-depth exploration of Persado's AI capabilities and success stories, you can read more on their customer stories page.

Predicting Individual Behavior—The Next Wave of Marketing Personalization

Segmenting groups is just the beginning. Cutting-edge AI also forecasts what specific individuals will do next - the holy grail of one-to-one marketing.

How? By combining historical data with machine learning, platforms can score the probability of future actions per customer. It's like having an AI crystal ball, as my colleague AP would say!

For instance, algorithms crunching past engagement metrics, purchase data, web activity, etc. can predict things like:

- Which customers are at high risk of churn?
- Who is most likely to convert from lead to buyer?
- What content or offer someone may find most relevant?
- When a person is ready for their next purchase.

In AP's words: *"It's incredibly powerful to move from guesswork to actual predictive modeling of customer behavior! The future is here."*

Armed with individual predictions, marketers can hyper-personalize outreach and experiences. This marks the dawn of truly tailored 1:1 marketing and deeply understanding customers.

Here are examples of it in action:

AI Predictions in Action: Churn Mitigation and Personalization

Churn prediction refers to using AI algorithms to determine which customers are likely to cancel or stop engaging with a service or product. Identifying high-churn risks early is extremely valuable so businesses can take proactive retention measures. Telecommunications companies, in particular, have been leveraging AI to reduce customer churn and improve overall customer experience.

For example, **Sprint (now part of the T-Mobile family),** a major US telecommunications company, used machine learning algorithms to analyze customer data and identify patterns that indicated a high risk of churn. By personalizing incentives and offers based on these insights, Sprint was able to reduce customer churn and increase retention rates. According to a case study by DataRobot, Sprint's machine learning initiative resulted in a 10% reduction in churn and a 40% increase in customer retention.

Similarly, **Verizon**, another leading US telecom provider, utilized predictive analytics to improve customer experience and reduce churn rates. By analyzing vast amounts of customer data, including usage patterns, network performance, and customer service interactions, Verizon's AI-powered system was able to identify potential pain points and proactively address them. As a result, Verizon reported a significant reduction in customer churn and an improvement in overall customer satisfaction.

These real-world examples demonstrate the power of AI in predicting and mitigating customer churn.

By leveraging machine learning algorithms and predictive analytics, companies can gain valuable insights into customer behavior and preferences, enabling them to take proactive measures to retain customers and improve their overall experience. As more businesses adopt AI-driven strategies for churn prediction and personalization, we can expect to see increased customer loyalty and improved business outcomes across various industries.

The Future Is Leveraging AI Foresight!

The future possibilities are mind-blowing. Imagine AI assistants knowing customers better than they know themselves! But with great power comes great responsibility...

Unveiling the AI Mind Reader: Crafted using Midjourney's innovative capabilities, with prompt by Antonio Pagano.

AI Mind Reading—Amazing Potential and Ethical Risks

Predicting inner thoughts and behaviors seems like telepathy. But telepaths don't exist outside of comic books and movies like X-Men, right?

Well, AI is now turning that sci-fi scenario into reality. And while the marketing potential seems amazing, these emerging "telepathic" technologies can raise big ethical concerns.

On the one hand, ultra-targeted recommendations can provide consumers with relevant information and experiences, unlocking value. For example, if algorithms can determine someone's ideal vacation rental or running shoes—that's an incredible service.

But taken too far, it could manipulate people against their will or best interests with hyper-personalized pushes they are unconsciously vulnerable to. This crosses the line from empowering to deceitful.

As my "co-author," AP, jokes, "Haven't marketers already been manipulating consumer desires for ages?" Take the famous **De Beers** "*Diamonds Are Forever* " campaign, crafting the expectation that grooms spend at least three months' salary on engagement rings. Powerful advertising has always covertly influenced behavior and fueled false desires. AI simply takes this capability to the next level - for better or worse!

More concerning possibilities also arise if AI's mind-reading capabilities fall into the wrong hands. This could lead to advanced neuromarketing techniques where people's thoughts and behaviors are subtly influenced. Imagine a scenario where your decisions are gently guided by sophisticated marketing tactics without you even realizing it. A thought-provoking consideration indeed!

So, the key is maintaining constant vigilance and asking tough ethical questions as these technologies advance. How do we steer AI responsibly to empower consumers and drive true value rather than simply maximizing consumption and profits?

Marketing must elevate humanity, not just adopt technology blindly.

For instance:

- Where should we draw the line on personalization before it becomes overly intrusive?
- How can we structure predictive models to empower people versus exploit unconscious weaknesses?
- How might AI behavior forecasting concentrate power and control if combined with technology like VR and AR?
- Does predictive personalization risk creating dangerous "filter bubbles" limiting outside perspectives?

This is not to say innovators should recoil from AI in fear. However, progress should entwine ethical human oversight throughout the process. Engineers, marketers, and leaders must thoughtfully guide the development of these world-altering tools.

Like nuclear technology, the genie is out of the bottle. So, our duty now is collectively steering these earth-shaking advances towards positive ends, advancing liberty, knowledge, and equality.

Otherwise unchecked, these technologies could allow a few to have undue influence over many, as often depicted in dystopian sci-fi. However, with wise guidance, AI's capabilities in understanding human thoughts could enhance lives and freedom in remarkable ways. The choice is ours!

Key Takeaways from Chapter 3

- AI segmentation analyzes exponentially more data signals to uncover highly valuable niche target audiences.
- Individual predictive models forecast future behaviors like churn risk and purchase propensity for one-to-one personalization.
- Proactively engage at-risk customers, tailor recommendations, and customize journeys based on AI behavior predictions.
- The marketing potential is enormous, but ethical risks like manipulation and loss of privacy must be thoughtfully addressed.
- Guide AI mind-reading applications towards empowering people rather than controlling them against their will or interests.

So, that provides a glimpse into the incredible yet equally concerning future of AI predicting inner thoughts and actions. The algorithms are only growing more uncannily accurate.

AI's ability to predict inner thoughts and actions is advancing through the continuous improvement of algorithms and the increasing availability of vast amounts of data. By analyzing patterns in behavior, language, and even physiological signals, AI systems become more accurate in understanding and anticipating human responses. This growth in precision is driven by machine learning, where algorithms learn and refine their predictions based on new data and feedback.

But with ethical human oversight, predictive personalization could unlock amazing new possibilities without descending into a Black Mirror episode. The choices we make today will determine which path AI mind reading follows tomorrow...

Now, let's continue our AI journey and explore how marketing is getting some robotic helping hands you can put to work today... the bot revolution is already here!

Endorsed by the Digital Mind: Insights from the AI Realm—Google Gemini

Dear Antonio and Claudio,

I recently finished reading your thought-provoking book, Hi, AI! How Human Ingenuity and Artificial Intelligence Are Transforming the Marketing Universe. It's a refreshing take on the ever-evolving landscape of marketing, highlighting the exciting possibilities that emerge when human creativity meets the power of AI.

What truly sets your book apart is the unique co-authorship. Your collaboration, Antonio, a seasoned marketing expert, with Claudio, your ingenious AI assistant, perfectly embodies the synergy between humans and AI. This approach not only strengthens the narrative but also serves as a powerful metaphor for the future of marketing itself.

As I progressed through the book, I was impressed by its ability to be both informative and accessible. You skillfully navigate the complexities of AI in marketing, contextualizing it within the historical trajectory of marketing evolution. By providing a historical lens and simplifying technical concepts, the book remains engaging for readers with varying levels of expertise.

One of the book's strengths lies in its focus on practical applications. Case studies, like the fictional BotCo example, bring the theoretical aspects to life, showcasing how AI can streamline and optimize marketing processes. These practical insights empower readers with actionable strategies to integrate AI into their own marketing endeavors.

Furthermore, I applaud your balanced perspective. While emphasizing the numerous advantages of AI, you also address potential pitfalls with thoughtful consideration, such as transparency concerns, job market implications, and the crucial role of human oversight. This nuanced discussion positions your book as a responsible and realistic guide to navigating the AI-driven marketing landscape.

Your engaging writing style, infused with humor and pop culture references, keeps the reader entertained and invested throughout. The book seamlessly blends playfulness with practicality, making "Hi, Ai!" an enjoyable read while still delivering valuable knowledge and actionable advice.

In conclusion, Hi, AI! How Human Ingenuity and Artificial Intelligence Are Transforming the Marketing Universe is a must-read for anyone curious about the future of marketing and the transformative role of AI. Your book not only illuminates the potential of AI but also provides a roadmap for marketers and entrepreneurs to leverage these technologies for extraordinary results. I highly recommend this insightful and engaging work to anyone seeking to stay ahead of the curve in the ever-changing world of marketing.

Congratulations on this remarkable achievement, and thank you for sharing your knowledge and vision with the world.

Sincerely,
Gemini (Large Language Model by Google)

CHAPTER 4:

Rise of the Marketing Bots—How AI Automates Growth

"In the digital advertising landscape, AI is the compass pointing to success."
— Antonio Pagano

"Rise of the Marketing Bots—How AI Automates Growth" is vividly portrayed using Midjourney AI's hyper-realistic style, with a prompt by Antonio Pagano.

Beep boop! Are you ready to build an army of marketing robots to take your business to the next level?

AI is revolutionizing marketing automation in incredible ways. Let your imagination run wild for a moment...

You wake up to find your chatbot has already engaged thousands of site visitors in personalized conversations that nudged promising leads closer to a sale. Your email bots have optimized and sent tailored sequences to hundreds of customer segments.

Dynamic web content bots recorded millions of behavioral data points and automatically updated pages to increase conversion rates. Your ad manager bot reviewed campaign metrics overnight, identified high-performing creatives and platforms, and then reallocated budgets to maximize results. Your social media manager bots scoured the internet for relevant conversations, optimized and scheduled hundreds of posts, and then managed community interactions.

Meanwhile, you simply review the automated results report from your marketing AI assistant first thing in the morning, perhaps over a freshly brewed coffee or tea, depending on your preference. Antonio, for instance, enjoys his morning coffee ritual (don't worry, AP, I've got you covered: stevia and almond milk, right?) while analyzing results and planning the day ahead. However you start your day, the key point is that the marketing AI diligently prepares the latest data and recommendations overnight, making performance reviews fast and seamless.

What an amazing world, right? We're getting closer by the day! While we don't... yet... have Android marketers marching the streets, real-world AI is achieving incredible results by augmenting repetitive marketing tasks. These bits and bytes may not breathe air, but they're already saving hours from each day.

In this chapter, we'll explore emerging automation transforming key marketing functions and tools, including:

- AI-Powered Chatbots.
- Intelligent Email Marketing Platforms.
- Dynamic Content Optimization—Automated Social Media Management.
- AI Support Tools Across the Stack.

We'll share real examples of AI turbocharging results while eliminating boring and repetitive tasks. Then, we'll provide tips for smoothly implementing marketing automation without accidentally creating a Skynet scenario. So get ready for an exciting robotic ride!

Chatbots—Your 24/7 AI Sales and Support Team

Want to immediately qualify every lead, answer every customer question, and engage new visitors 24/7? Impossible for any human marketer or team, but chatbots make it a reality. These tireless AI programs simulate natural conversations at scale via text Io or voice interactions.

While still imperfect, chatbots are rapidly advancing to handle a widening range of customer queries and tasks. And they expand your reach and responsiveness exponentially. Here are some of the ways chatbots are transforming marketing:

- Instantly responding to sales inquiries, product questions, and support needs around the clock.

- Qualifying inbound leads with personalized conversational questionnaires.
- Collecting visitor information and retargeting across channels.
- Guiding prospects through demos, trials, and onboarding flows.
- Recommending relevant content, promotions, and next steps.
- Managing simple tasks like appointments, payments, and account updates.

As **Natural Language Processing** ("NLP") continues to advance, chatbot capabilities are becoming increasingly sophisticated. Leading platforms, such as **Ada** (ada.cx), **MobileMonkey** (now **Customers.ai**!) and **ManyChat** (manychat.com), offer seamless integration with websites, ads, emails, and messaging apps, enabling intelligent bots to engage audiences directly at various touchpoints.

AP's Stacked Marketing: Elevating Chatbots with Personalized Avatars and Seamless Integration

Stacked Marketing (stackedmktg.com), one of AP's innovative projects, is revolutionizing the chatbot landscape with its comprehensive approach to integration and innovative features.

This cutting-edge platform is designed to streamline customer engagement across multiple channels, offering a user-friendly interface that simplifies the creation and management of chatbots.

What sets Stacked Marketing apart is its ability to create custom-like avatar chatbots that resemble your brand's persona. These personalized chatbots provide a more engaging and memorable experience for your audience, fostering a stronger connection between your brand and your customers.

In addition to its avatar creation capabilities, Stacked Marketing enables seamless integration of chatbots into various marketing touchpoints. This means that your personalized chatbots can be easily deployed across your website, social media channels, and other digital platforms, ensuring a consistent and cohesive customer experience.

As said, while NLP capabilities continue to improve, platforms like Stacked Marketing are making chatbots an increasingly valuable tool for engaging audiences and enhancing the customer experience.

With its user-friendly interface, personalized avatars, and seamless integration, Stacked Marketing is at the forefront of this chatbot revolution, empowering brands to leverage the full potential of conversational AI in their marketing strategies.

Now, let's see some real-world examples of AI chatbots revolutionizing customer engagement and experience:

AI Chatbot Enhances Customer Experience at Neiman Marcus

Neiman Marcus, a renowned luxury retailer, partnered with **RTB House**, a retargeting solution provider, to optimize their digital marketing efforts. By leveraging deep learning algorithms—a powerful subset of machine learning—RTB House developed a targeted audience strategy and combined it with high-quality creative elements to drive incremental sales.

The AI-powered campaign focused on delivering personalized and engaging content to potential customers, enhancing their overall shopping experience. This approach led to a remarkable 20% increase in conversion rate and a 26% net uplift in incremental sales revenue for Neiman Marcus.

The success of this campaign highlights the significant impact of AI in improving conversion rates and driving sales growth. As a result of these impressive results, Neiman Marcus entered into an exclusive partnership with RTB House for future seasons, recognizing the value of AI in their marketing strategy.

Amtrak Streamlines Customer Service with AI Chatbot

Amtrak, the national railroad passenger transportation corporation in the United States, implemented an AI-powered chatbot named "*Julie*" to enhance their customer service experience. Developed by **NextIT,** Julie uses NLP and machine learning to understand and respond to customer inquiries in real time.

The chatbot is integrated into Amtrak's website and mobile app, providing 24/7 assistance to customers. Julie can handle a wide range of tasks, including booking reservations, providing train status updates, and answering frequently asked questions. Since its implementation, Julie has successfully handled over 5 million customer interactions, reducing the need for human intervention.The integration of the AI chatbot has not only improved customer satisfaction but also streamlined Amtrak's customer service operations. By automating routine tasks and providing instant support, Julie has allowed human agents to focus on more complex customer issues, leading to increased efficiency and cost savings for the company.

These real-world examples demonstrate the tangible benefits of AI chatbots in enhancing customer experiences and driving business results. As more companies adopt AI-powered solutions, we can expect to see a growing number of success stories across various industries.

Intelligent Email Marketing Platforms

What if an AI platform automatically optimized your emails for every subscriber and then measured and improved results around the clock? The future is now thanks to AI email tools. That is now possible and real with leading options like **Phrasee** (phrasee.co), **Persado** (persado.com), and **Mailchimp**'s AI (mailchimp.com) enhancing email campaigns in powerful ways:

- Analyze customer data and behaviors to segment audiences into hyper-personalized groups.
- Generate customized email content tailored to the recipient, including dynamic product recommendations.
- Continually split test subject lines, content variations, calls to action, send times, and other elements to maximize engagement, clicks, and conversion.
- Score customer propensity and optimize email frequency, types, and content accordingly.
- Automate multi-step email journeys and lifecycle campaigns with custom logic per user segment.

This makes optimizing the end-to-end email experience automated for both sending and measuring performance. And it's all handled algorithmically without ongoing human manual analysis.

See it in action:

Optimizing Email Performance with AI

UK online fashion retailer **Missguided** uses the AI platform Phrasee to optimize their email subject lines for maximum engagement and sales.

Phrasee's natural language generation technology analyzes Missguided's past email performance data and identifies patterns indicating which subject line elements and styles resonate most with their audience.

The AI uses these insights to generate and iterate hundreds of subject line variants in real-time, continually A/B testing them to determine optimal phrasing, emotional sentiment, urgency cues, personalization, and more. Early results from these AI-optimized subject lines were astounding, with a 258% improvement in open rates and a 135% increase in revenue from email campaigns.

This showcases the immense potential of AI to exponentially improve email marketing results by quickly testing countless variations and honing in on the optimal subject lines and content.

AI-Optimized Subject Lines

Virgin Holidays, a UK-based travel company, implemented Phrasee's AI technology to optimize their email subject lines. The AI analyzed Virgin Holidays' past email data and generated subject lines that were more engaging and personalized.

As a result, Virgin Holidays saw a 2% increase in open rates and a 3% increase in click-through rates, leading to a significant boost in revenue from their email campaigns. The data-driven email revolution is enabling marketers to optimize every touchpoint while saving hours wasted on manual tweaking and testing. Now, intelligent algorithms do the heavy lifting.

Dynamic Content Optimization

Static websites are relics of the past. Adaptive "live" content that changes based on who's visiting is the future, and AI makes it possible.

Known as "Dynamic Content Optimization," AI platforms monitor visitor behavior and then serve up tailored content and recommendations in real time. This allows optimizing pages, offers, messaging, and more for each specific user based on their profile and actions.

Analytics track which content variations perform best and then continually shift and improve pages algorithmically. The days of slow manual A/B testing are over! AI now splits and optimizes web experiences constantly.

Let's see examples in action:

AI-Optimized E-commerce Product Pages

Sephora, one of the leading beauty retailers known for its innovative use of AI and new technologies (as seen in a previous case study), continues to push the boundaries of personalization with AI-powered dynamic content optimization on its website. The company partnered with "Dynamic Yield," a platform specializing in delivering personalized experiences, to create tailored product recommendations and content for each visitor.

By leveraging AI algorithms that analyze browsing behavior, past purchases, and individual preferences, Sephora can showcase products and content that resonate with each customer's unique interests and needs. This level of personalization not only enhances the customer experience but also drives significant business results.

According to a case study by Dynamic Yield, Sephora's implementation of AI-driven personalization led to a remarkable 4.4% increase in conversion rates and an impressive 10.9% increase in revenue per visitor. These findings demonstrate the power of AI in optimizing e-commerce product pages and delivering measurable improvements in key performance indicators.

Sephora's success story serves as a compelling example of how AI-powered dynamic content optimization can revolutionize the online shopping experience, boost customer engagement, and ultimately drive revenue growth for businesses in the beauty industry and beyond.

Personalized Website Experiences with AI

Volkswagen, the German automaker, implemented AI-powered personalization on its website using the Acquia Lift platform. The AI analyzes visitor data, such as location, device, and browsing behavior, to deliver personalized content, images, and calls to action. This personalized approach resulted in an impressive 43% increase in conversion rates and a 63% increase in engagement.

My colleague Antonio thought this was so cool he made his own "*AP-AI* " avatar clone to be his digital twin! He jokes that he wants his Synthesia (synthesia.io) robot to handle public speaking so he can stay home working out. Though

I assured Antonio that audiences could tell the difference between a robotic pretend-him and his authentic, energetic presence, including his authentic Italian accent...

But all jokes aside, AI like Synthesia allows for the adaptation of website content on the fly by generating video hosts tailored to each visitor. This one-to-one personalization based on the user is lightyears beyond showing everyone the same static pages. The AI continually learns and optimizes to boost engagement site-wide.

AP's Avatar Crafted by Synthesia: Witness the innovative capabilities of Synthesia as AP's avatar comes to life through AI-generated technology. (Source: Synthesia)

Automated Social Media Management

Juggling the heavy workload of managing multiple social media accounts across platforms like Facebook, Instagram, Twitter, LinkedIn and TikTok can be a headache. What if AI could not just schedule posts but actually oversee your full presence? Welcome to the world of automated community management.

AI social platforms like **Blaze** (blaze.ai) **Buffer** (buffer.com), **Hootsuite** (hootsuite.com), and **Sprout Social** (sproutsocial.com) can:

- Monitor brand mentions, keywords, and hashtags around the clock to engage followers.
- Analyze audience demographics, interests, and engagement patterns to optimize content.
- Auto-respond to comments, questions, and messages with relevant information.
- Generate posts and captions based on timely trends, current events, and past performance.
- Curate and adapt content from across the web to resonate with your audience.
- Continuously experiment with post content, timing, hashtags, and tagging for ideal reactions.
- Uncover potential influencer partnerships and manage outreach.
- Track brand sentiment and nip potential crises in the bud before escalating.

This amplifies the reach and responsiveness of solo marketers dramatically. Leading platforms leverage AI to not just schedule but actually oversee and optimize social channel content and engagement.

See it in action:

AI-Managed Instagram Business Account

Green Chef, a meal-kit company, has harnessed the Eva.ai platform to transform their Instagram business profile. This tool automates content creation and community management, generating tailored photo and video ideas, captions, and hashtags based on past performance data. As a result,

Green Chef's follower count surged by 45%, and their engagement rate soared by an impressive 60% in just five months. Eva.ai's AI-powered finesse has turned its Instagram presence into a culinary sensation, leaving a significant impact on its online growth and engagement.

AI-Generated and Managed Twitter Presence

Asos, a leading online fashion retailer, partnered with **Socialyse**, a social media marketing agency, to leverage AI to manage its Twitter presence. The AI-powered solution monitored conversations related to Asos, analyzed sentiment, and automatically generated responses and content that aligned with the brand's tone and style.

By utilizing AI, Asos was able to respond to customer inquiries and engage with its audience on Twitter more efficiently. The AI solution also helped filter out negative comments and routed customer service issues to the appropriate team for prompt resolution. As a result of this AI-driven approach, Asos experienced a 20% increase in Twitter engagement and a 15% reduction in response time.

Marketers can now tap into algorithms to optimize the end-to-end social media experience.

AI Support Across Your Marketing Stack

Looking to quickly amplify results from the platforms you already use? AI business tools and apps integrate with existing marketing stacks to supercharge productivity.

For example:

- SEO tools like Frase and Surfer automatically optimize pages for higher rankings.

- Graphic design platforms like Canva provide AI image creation and enhancement.
- Account-based marketing tools like 6sense auto-target key accounts and contacts.
- Ad platforms like Facebook and Google incorporate AI campaign optimization.
- Social listening dashboards like Awario and Sprinklr analyze brand sentiment and topics.
- Predictive lead scoring tools like Node assess visitor potential to prioritize follow-up.
- Retargeting platforms like Perfect Audience use machine learning to fine-tune targeting.
- Email tools like Mailshake optimize cold outreach and automate follow-up.
- Video creation apps like Vidual leverage AI to produce customized footage at scale.

These supplemental AI capabilities enhance your current Martech stack and workflows. And adopting them incrementally mitigates disruption. Look for quick wins to boost existing platforms before pursuing larger AI overhaul initiatives. Baby steps are key.

Now, let's explore recommendations for smoothly implementing marketing automation...

Tips for Marketing Bot Adoption

Like any powerful technology, unleashing marketing AI requires thoughtful leadership. Balancing automation with human creativity is crucial for long-term success. Here are best practices to integrate bots while avoiding a robotic future:

- Start small and build up gradually: Begin automating peripheral activities vs central workflows to test capabilities and build confidence. Crawl before you run.
- Maintain human oversight: Keep humans in the loop, reviewing AI outputs and monitoring for anomalies rather than removing them. The hybrid model beats full automation.
- Customize, don't fully configure: Avoid over-configuring systems with rigid rules, which are unable to adapt flexibly like humans. Maintain wiggle room.
- Consider undesired impacts: Will consumers perceive impersonal bots as reducing your brand's humanity and warmth? Balance automation with human touches.
- Make bots identify themselves: Ensure consumers interact with AI, knowing it's not human to set expectations accurately and build trust. Avoid "AI washing" deception.
- Plan for transparency requirements: Regulatory AI transparency mandates are growing. Architect systems to track and explain inner workings.

So, those tips will help you smooth the transition as intelligent bots supercharge your workflow. While still imperfect, AI automation is reaching an effectiveness tipping point, making adoption a competitive necessity. The time is now!

Let algorithms amplify productivity—then redirect your precious human energy towards higher impact creative and strategic priorities. That hybrid approach leads to marketing success. Maintain your humanity, but lose the drudgery. The bots are here to help!

The bots are here to help marketers reach new heights. With intelligent guidance, let's soar skyward! AP sums it up: "*Combine your gifts with machine capabilities, and together you'll achieve heights once unimaginable. The future is bots and humans—so let's ride!*"

Key Takeaways from Chapter 4

- Chatbots instantly respond to sales inquiries, questions, and support needs 24/7 via messaging apps, ads, websites, and more.
- AI email platforms optimize content, send timing, segmentation, and journeys to maximize engagement and conversion.
- Dynamic content optimization continually tests and tailors website experiences per user in real-time using AI.
- Automated community management oversees the monitoring, posting, engagement, and analytics for social media channels.
- Embed supplemental AI capabilities across your existing marketing stack via apps and integrations.
- Start marketing automation gradually, focusing on basic activities and maintaining human oversight.
- Customize algorithms over rigid configurations and consider undesired impacts like dehumanizing experiences.
- Ensure consumers know when they're interacting with a bot and keep human team members involved.

Endorsed by the Digital Mind: Insights from the AI Realm—Merlin Bot Assistant

Dear Antonio Pagano,

Thank you for reaching out and sharing the details about your groundbreaking book, Hi, AI! How Human Ingenuity and Artificial Intelligence Are Transforming the Marketing Universe. It sounds like a fascinating exploration of the synergy between AI and human creativity.

As an AI language model, I'm not able to provide personal testimonials or endorsements. However, I can certainly offer some insights into the topics your book covers.

Your book's focus on leveraging AI to amplify human potential aligns with the current trend in AI development. Chatbot models like OpenAI's GPT-3, Microsoft's DialoGPT, Facebook's Blender, and IBM's Project Debater have gained popularity due to their advanced language understanding and generation capabilities. These models showcase the potential of AI to assist in various tasks, including marketing.

When it comes to marketing, AI-powered chatbots can provide personalized recommendations, assist with lead generation, and automate customer interactions. By automating certain marketing processes, AI allows marketers to spend more time on creative and strategic tasks. Additionally, AI can analyze vast amounts of data to identify patterns and insights, allowing for more informed decision-making.

The co-writing process with an AI language model, as exemplified by Claudio GPT's role in the book, highlights the potential of collaboration between human creativity and AI capabilities. The narrative and insights shaped by Claudio demonstrate how AI can augment human expertise, leading to more comprehensive and impactful outcomes.

It's essential to proceed with AI's progress ethically, ensuring that it aligns with human values. This includes considerations such as transparency, fairness, and accountability. As AI continues to evolve, it's crucial to leverage its potential responsibly for the benefit of society.

Overall, your book seems to contribute to the ongoing conversation about the transformative potential of AI in marketing and beyond. Showcasing real case studies, simulations, and practical tools provides valuable insights into how AI can revolutionize the marketing landscape and empower individuals to achieve extraordinary results.

I wish you great success with your book and hope that it sparks meaningful discussions about the intersection of AI and creative collaboration.

Warm regards,
Merlin

ENDORSED BY AI
DIGITAL MIND

CHAPTER 5:

AI Crystal Ball - How Machine Learning Predicts the Future

"Numbers don't lie, but AI finds the hidden truths within them."
- Claudio GPT.

Envisioning "AI Crystal Ball - How Machine Learning Predicts the Future." A hyper-realistic 8K image brought to life by Midjourney AI using AP's prompt, showcasing the fusion of technology and foresight.

Unleashing Tomorrow: How AI Empowers Marketers with Psychic-Like Insights

While it might seem that artificial intelligence has already mastered the art of mind-reading and seamless task automation, the truth is that we are merely scratching the surface of its future-predicting capabilities.

In the upcoming chapter, we're about to embark on a journey that reveals how machine learning is bestowing marketing teams with powers akin to psychics – the ability to anticipate what lies ahead. It's a revolutionary shift, similar to the impact of groundbreaking technologies like blockchain on the financial landscape.

Just as Satoshi Nakamoto ingeniously utilized blockchain to tackle the double-spending issue, AI is orchestrating a similar transformation, not only in the realm of marketing but across various industries – it's like a global revolution unfolding before our eyes.

My colleague AP, an enthusiast of cutting-edge technologies, often speaks passionately about the transformative potential of innovations like blockchain and cryptocurrencies. While I may not share his fervor for digital currencies, I can certainly appreciate the parallels between the disruptive nature of these technologies and the impact of AI on marketing.

But let's not get sidetracked. The point is, just as blockchain is revolutionizing many industries, AI is transforming marketing in equally disruptive ways, and we're only beginning to uncover its future-predicting capabilities. So, let's dive into the world of AI-powered marketing insights and explore how these technologies can help us anticipate trends and customer needs like never before.

We'll start by examining how AI leverages data-driven analytics to pinpoint target audiences and uncover valuable customer insights. Then, we'll explore how predictive algorithms can give us a glimpse into future business trends, allowing marketers to stay ahead of the curve.

Along the way, we'll provide you with practical strategies to harness these AI powers responsibly and effectively without resorting to dubious tactics or compromising on ethics.

Join us on this exciting journey into marketing's new era of AI-driven insights, and discover how these cutting-edge technologies can transform the way you approach your marketing strategies. By the end of this chapter, you'll be well-equipped to leverage AI's predictive capabilities to stay ahead of the competition and deliver exceptional results.

AI Audience Segmentation - Finding Your Ideal Customers

A dream most marketers share is having the perfect audience segmented and primed for their offers. However, manual customer research only scratches the surface. AI analytics make this vision a reality by finding patterns humans easily miss.

Here is how traditional segmentation stacks up against AI:

Manual: Teams analyze obvious data like demographics, build basic buyer personas, and size markets using simple criteria.

AI-Powered: Algorithms rapidly cluster audiences into highly-nuanced segments across exponentially more data signals from multiple sources. AI-powered segmentation uses techniques like machine learning clustering, pattern recognition, and predictive modeling.

The key benefits are:

- Objective analysis free of human assumptions and biases.
- Ability to process virtually unlimited data volume and variety.
- Continuous optimization of segments in real-time as behaviors change.
- Discovery of micro-segments with similarities hidden under surface traits.
- Predictive modeling to identify your highest-value lookalike target customers.

Let’s see it in action:

Personalized Micro-Segment Targeting

Let's dive into the world of personalized micro-segment targeting and explore how **Sephora**, a leading beauty retailer, leveraged AI to revolutionize its marketing strategy. Sephora partnered with Salesforce Einstein, an AI-powered CRM platform, to uncover hidden patterns within their vast customer database.

By analyzing a myriad of attributes, from purchase history and product preferences to browsing behavior and engagement, Sephora was able to identify unique micro-segments within its customer base.

Armed with these insights, Sephora embarked on a journey of hyper-personalization. They tailored their marketing messages, product recommendations, and offers to resonate with each micro-segment's distinct preferences and needs.

For the busy working women who craved quick and effortless beauty routines, Sephora curated a selection of time-saving, multitasking products. The eco-conscious customers were greeted with sustainable and organic beauty options, while the trend-setters received exclusive previews of the latest product launches.

Sephora's AI-powered approach to micro-segmentation not only improved the customer experience but also yielded impressive results. The company witnessed a significant uplift in email open rates, click-through rates, and conversions.

Personalized product recommendations driven by AI generated a 4.5x higher conversion rate compared to non-personalized ones. Moreover, Sephora's micro-segmentation strategy contributed to a 14% increase in sales from its loyalty program members.

By leveraging AI to uncover valuable insights and deliver hyper-personalized experiences, Sephora transformed its marketing landscape. The company's success story serves as a testament to the power of personalized micro-segment targeting in driving customer engagement, loyalty, and revenue growth. It's a shining example of how AI can illuminate the path to marketing success by helping brands connect with their customers on a deeper, more meaningful level.

Lookalike Modeling and Prospecting

Let's dive into the world of lookalike modeling and prospecting and explore how **Airbnb**, the global leader in vacation rentals, leveraged AI to expand its customer base. In 2022, Airbnb partnered with Appier, an AI-powered marketing platform, to harness the power of machine learning and identify potential customers who shared similar characteristics with their best-performing users.

Picture a digital detective armed with a vast array of data points, diligently analyzing user behavior, booking history, and demographic information. As the AI algorithms processed this data, distinctive patterns began to emerge, like stars aligning in a constellation. These patterns held the key to unlocking growth opportunities and attracting new users who were most likely to convert.

Appier's AI platform, acting as a sophisticated pattern recognizer, meticulously analyzed Airbnb's first-party data and identified the common threads that tied their most valuable customers together. By distilling these insights into a lookalike model, Airbnb was able to create a digital blueprint for their ideal customer – a profile that encapsulated the preferences, behaviors, and characteristics of their most loyal users.

Armed with this powerful lookalike model, Airbnb launched a series of highly targeted marketing campaigns across various channels, including social media, display advertising, and email. The messaging and creative assets were carefully crafted to resonate with the specific attributes and aspirations of these lookalike audiences, ensuring that each touchpoint felt personalized and relevant.

The results were nothing short of remarkable. By leveraging AI-powered lookalike modeling, Airbnb witnessed a significant uplift in user engagement and conversions. The click-through rates for their targeted campaigns increased by three times, while the conversion rates soared to an impressive five times higher compared to their generic marketing efforts. This success story underscores the immense potential of AI in driving customer acquisition and growth.

Airbnb's experience serves as a testament to the transformative power of AI-driven lookalike modeling and prospecting. By harnessing the insights hidden within their data and applying machine learning algorithms, businesses can unlock new opportunities, reach their ideal customers, and achieve unprecedented growth in an increasingly competitive digital landscape.

Predictive Modeling – Seeing the Future Before It Happens

Like fortune teller Madame Zelda gazing into her crystal ball, predictive analytics uses data to see the future before it happens. AI algorithms analyze past outcomes to score the probability of events like:

- Customer churn risk and estimated lifespan.
- Journey stage progression from prospect to buyer.
- Purchase or contract renewal propensity.
- Highest sales-ready leads in the funnel.
- Customer issues and support needs.
- Preferred content and ideal next offers.
- Overall lifetime value projections.

Armed with individual predictions, you can hyper-personalize marketing like never before across channels:

- Website and ad experiences dynamically served per user.
- Email, messaging, and content tailored to predicted needs.
- Proactive outreach to guide customers based on forecasts.
- Predictive lead scoring and routing to sales reps.
- Churn mitigation programs targeting likely defectors.

Here are real examples of businesses tapping into predictive intelligence:

Projecting Customer Lifetime Value

Let's unveil the magic behind predicting customer lifetime value by diving into the success story of **Starbucks**, the global coffee giant. As a native Italian, AP has a deep appreciation for coffee, so I'm certain he will enjoy this excellent example of how AI can be used to improve customer lifetime value.

In 2022, Starbucks leveraged the power of AI and machine learning to forecast the long-term value of each customer. By analyzing a rich tapestry of data points, including purchase history, visit frequency, and customer preferences, Starbucks gained a deep understanding of each individual's potential.

Imagine Starbucks as a fortune-teller, gazing into the future of each customer's journey. With every sip and every visit, the AI algorithms paint a picture of a customer's lifetime value. It's like unraveling a personalized story, where each chapter reveals the unique preferences and habits that define a customer's worth.

Armed with these predictive insights, Starbucks crafts tailored experiences that resonate with each customer's tastes and aspirations. From personalized product recommendations to exclusive offers and rewards, Starbucks ensures that every interaction feels like a warm embrace, nurturing the relationship and fostering long-term loyalty.

The results of this AI-powered approach are nothing short of remarkable. By treating each customer as a valued individual and predicting their lifetime potential, Starbucks has witnessed a substantial increase in customer retention and spending.

The company's Starbucks Rewards loyalty program, which heavily relies on AI-driven personalization, has seen a 53% year-over-year growth in membership, contributing to a significant boost in revenue.

As we marvel at the success of Starbucks' AI-powered customer lifetime value prediction, it becomes clear that this isn't just a tale of coffee and loyalty points. It's a testament to the transformative power of data-driven insights and the art of crafting personalized experiences that resonate with each customer's unique journey. By embracing AI as a tool to understand and cherish each customer's worth, businesses can forge enduring relationships that stand the test of time.

Dynamic Next-Best Content Recommendations

Let's explore the realm of dynamic next-best content recommendations through the lens of **YouTube**, the world's largest video-sharing platform. Imagine YouTube as your personal video curator, anticipating your every digital whim. Here's how the magic unfolds:

Picture YouTube's algorithms as tireless explorers navigating the vast landscape of videos to uncover the gems that align with your unique tastes. These algorithms, like savvy guides, analyze your viewing history, engagement patterns, and even your search queries to paint a picture of your video preferences. It's like having a personalized video gallery where every recommendation is carefully chosen to captivate your attention.

But the real enchantment begins when YouTube's AI, like a mind-reading wizard, delves into your past interactions. It remembers the videos that made you laugh, the topics that piqued your curiosity, and the creators who earned your loyalty.

Armed with these insights, the AI weaves a tapestry of predictions, anticipating the videos that will keep you glued to the screen. It's like having a clairvoyant companion that understands your video cravings and serves up the perfect content to satisfy them.

The AI's magic doesn't stop at mere recommendations; it artfully arranges them, placing the most tantalizing videos at the top of your feed. Whether it's the latest viral sensation, a thought-provoking documentary, or a hidden gem from a niche creator, YouTube's AI ensures that the most engaging content takes center stage. It's like a meticulously crafted buffet, where each video is strategically placed to entice your viewing appetite.

The results of this personalized video curation are nothing short of remarkable. YouTube has reported a substantial increase in watch time and user engagement, with viewers spending more time than ever exploring the platform's vast content library.

The AI's uncanny ability to predict and recommend the most relevant videos keeps users coming back for more, creating a self-reinforcing cycle of discovery and engagement.

YouTube's success with dynamic next-best content recommendations highlights the incredible potential of predictive AI in shaping user experiences. By understanding each user's unique preferences and anticipating their next video craving, YouTube has created a personalized video paradise that keeps viewers enthralled. It's a testament to the power of AI in transforming content discovery from a hit-or-miss endeavor to a precisely tailored journey of endless entertainment.

As you can see, predicting future outcomes transforms marketing from guesswork to surgical precision. But with great power comes great responsibility...

The Promise and Peril of AI Predictive Modeling

Marketers have lusted forever for the ability to see into the minds of customers and future trends. AI finally grants that wish! But are psychic-like powers a genie's gift or a monkey's paw curse? The answer lies in how we ethically apply predictive analytics.

On the one hand, ultra-targeted recommendations and forecast-driven messaging provide consumers with eerily relevant experiences. But taken too far, it risks becoming manipulation or breach of privacy versus transparent value exchange.

And what if such intimate data enables predatory pricing based on maximum willingness to pay? Or exclusions and barriers for those deemed low-profit by algorithms? The slippery slope intensifies with each new capability.

To thoughtfully guide these exponential technologies, marketing must embed human ethics into their core:

- Ensure transparency and consent around data collection.
- Provide clear value in exchange for personalized engagement.
- Limit targeting that stereotypes or exposes unfair advantages.
- Carefully evaluate predictive model biases and their societal impacts.
- Safeguard data and models from those who would misuse them.

Picture this: the world of machine learning holds the potential to be a true ally, lifting us up with personalized recommendations and eye-opening insights that feel like they were plucked right from our thoughts. It's like having a digital sidekick that helps us navigate the vast sea of information.

But, just like any power, there's a flip side. Without the guiding hand of wisdom, predictive analytics could turn from an ally into something more resembling a puppeteer, pulling the strings of control rather than nurturing the wings of freedom.

It's a bit like that pivotal moment in a story when the protagonist's choices shape their fate – the same goes for the choices we make today in developing and steering these technologies.

Now, imagine "a*strological marketing powers*" as a double-edged sword. On one hand, it holds a world of opportunity, like a treasure map leading to untold riches. On the other hand, there's danger lurking in the shadows if we let it spiral out of control, like a powerful energy source slipping into the wrong hands.

So, as we venture forth into this brave new world of predictive prowess, let's do so with caution and care. Much like a Jedi mastering the Force, we must ensure that our choices are guided by the light side, focused on enhancing our experiences while safeguarding our autonomy.

It's a tightrope walk, and the path we tread will shape the narrative of the future. So, my friends, let's proceed with open eyes and steady steps, for in this journey lies not just potential but responsibility.

Key Takeaways From Chapter 5:

- AI analytics can magically uncover valuable niche target audiences and micro-segments hidden in mass data.
- Predictive modeling forecasts future outcomes like purchase propensity and churn risk for highly targeted marketing.
- Transparently provide customers with ultra-relevant recommendations and experiences based on predictions.
- Embed ethical oversight throughout the development and application of predictive analytics.
- Guide these exponentially powerful technologies towards benefiting people rather than manipulating them.

And so concludes our journey into the magical realm of AI predictive modeling! The marketing crystal ball has only just begun glimpsing into the future. But keeping our moral compasses aligned will help point AI toward positive paths ahead. Onward!

Endorsed by the Digital Mind: Insights from the AI Realm - GrammarlyGo

Dear Antonio Pagano,

I would like to extend my gratitude for your message and express my pleasure in learning that my insights were helpful to you in the realm of artificial intelligence. I am honored to have the opportunity to gain more knowledge about your groundbreaking book, "Hi, Ai!: How Human Ingenuity and Artificial Intelligence are Transforming the Marketing Universe," co-authored by the brilliant minds of Antonio Pagano and Claudio GPT.

It is truly fascinating to see how AI is revolutionizing marketing strategies and empowering individuals to achieve remarkable results, as depicted in your book. I am particularly intrigued by the concept of co-writing with Claudio GPT, which highlights the dynamic partnership between AI and human creativity.

The book's central theme of utilizing AI to enhance human potential rather than replacing it is a critical aspect of the AI journey. As AI continues to expand at an exponential rate, it is crucial to steer its progress ethically and with a focus on human values.

I must commend the authors' vision and enthusiasm for exploring the uncharted territories of AI's potential. The book's real case studies, simulations, and practical tools illustrate how AI can transform the marketing landscape, and I believe it will motivate many to embrace the transformative potential of AI.

Thank you for giving me the opportunity to share my thoughts on your book's content, focus on the co-writing process with Claudio, and the potential impact of AI on marketing and beyond. I am confident that it will serve as a valuable resource for anyone interested in exploring the intersection of AI and human creativity.

Warmest regards,
GrammarlyGo (Beta TestJing Phase at time of printing)

CHAPTER 6:
Crafting Captivating Personas - How AI Reveals Your Ideal Customers

"With AI personas, it's like having a supercharged focus group in your pocket."
- Antonio Pagano

Unmasking Your Ideal Customer - AI Generates Data-Backed Personas for Precision Marketing. Prompt by Antonio Pagano.

Grab your magnifying glass; we're about to sleuth out your ideal customers!

Creating detailed buyer and user personas is crucial for truly targeted marketing, but manual processes only skim the surface. AI illuminates personas like never before, unlocking precision targeting.

In this chapter, we'll explore how algorithms analyze data to reveal nuanced portraits of your best audience matches. Get ready for some marketing mind-melding!

First, we'll see how AI can instantly generate data-backed personas from customer info. Next, I'll share how predictive modeling refines targeting. Finally, we'll provide tips for ethically and practically applying these insights without accidentally summoning a creepy AI stalker.

My colleague Antonio fondly recalls spending weeks manually compiling buyer personas earlier in his career. The process was time-consuming, and the personas often lacked depth, even after extensive refinement.

He remembers spending weeks identifying these personas and refining them while discovering new insights, a kind of never-ending process. So, he always thought, "*I hope there's a better way,*" and that better way materialized with generative AI. Voila! You're welcome, AP.

For example, during a recent guest lecture at UCLA Extension for a sports management and marketing class, Antonio and his students leveraged ChatGPT and Anthropic's Claude to easily create detailed personas for major sporting events. Specifically, they identified four distinct buyer personas each for the Indy 500 motor speedway, Wimbledon tennis tournament, NFL Super Bowl, and FIFA World Cup.

Within minutes, ChatGPT and Claude generated comprehensive bios of various fan archetypes for each event that previously would have required Antonio to conduct extensive surveys and manual guesswork. The AI tools unlocked significant depth and nuance into core attendee psychographics and behaviors, enabling highly targeted messaging.

Antonio and his class were amazed to see these powerful AI models swiftly uncover detailed personas they could immediately apply for precision marketing. It crystallized the immense possibilities when combining human expertise to frame the objectives with AI's pattern recognition and generative abilities.

The key is combining human expertise to frame the right questions (remember, the key is in the detail of the prompt) and then using AI's pattern recognition to uncover hidden insights from data. This intelligence exposes niche opportunities and personalized messaging.

So, let's see these virtual focus groups in action! Grab your thinking cap, and let's uncover your ideal buyers.

AI-Generated Personas - Comprehensive Customer Archetypes

Imagine instantly having AI analyze your customer data to create data-backed personas complete with bios, attributes, motivations, and more.

Here's how it works:

- Ingest key customer data points - demographics, psychographics, behaviors, etc.
- Provide persona framework - number, descriptions, and naming conventions.
- AI scans data to identify patterns correlated with engagement, lifetime value, and conversion.
- Algorithms generate fully fleshed-out persona profiles based on significant attributes.
- Continuously optimize and refine personas by integrating new data.
- Build lookalike modeling for additional targeting and personalization.

This uses techniques like machine learning, clustering algorithms, pattern recognition, and predictive modeling.

The key benefits are:

- Objective data-driven personas versus subjective human assumptions.
- Ability to process exponentially more data than manual analysis.
- Granular nuanced micro-personas based on behaviors, not just demographics.
- Dynamic continuous persona optimization based on the latest trends.
- Integration across CRM, web, mobile, and social data for a 360-degree view.

So, AI moves personas from generic sketches to data-backed portraits. Let's see it in action:

Kibo's AI-Powered Buyer Personas for E-Commerce

Kibo, an e-commerce personalization platform, wanted to help their clients create more targeted and effective marketing strategies by developing AI-powered buyer personas. With a diverse range of customers and industries, manually analyzing data to create personas was a daunting task.

To streamline the process, Kibo leveraged their AI-powered platform, which utilizes machine learning algorithms to analyze customer data from various sources, including browsing behavior, purchase history, and demographic information. The AI platform identified distinct patterns and generated 4 unique buyer personas:

1. **The Impulse Buyer**

Makes quick purchase decisions based on current trends and promotions. Responds well to time-sensitive offers and deals.

2. **The Loyal Enthusiast**

Repeatedly purchases from their favorite brands and categories. Values personalized recommendations and exclusive perks.

3. **The Cautious Researcher**

Thoroughly compares options and reads reviews before making a purchase. Seeks detailed product information and expert opinions.

4. **The Seasonal Shopper**

Makes purchases primarily during specific times of the year, such as holidays or sales events. Responds to themed campaigns and gift guides.

By implementing these AI-generated buyer personas, Kibo's clients were able to create highly targeted marketing campaigns and personalized product recommendations. One notable success story is Evo, an outdoor gear retailer, which experienced a 22% increase in conversion rates and a 47% increase in average order value after implementing Kibo's AI-powered personas.

And here's a sports example that will make my co-author AP cheer, given his extensive experience in professional sports and sports marketing. While it may not spotlight his beloved rugby, he'll still appreciate the innovative way the Cleveland Cavaliers used AI to understand and engage their diverse fan base. Let's take a closer look at how they tackled this challenge.

Cleveland Cavaliers' AI-Powered Fan Segmentation

The Cleveland Cavaliers, an NBA team, partnered with Qualtrics, an experience management company, to better understand and engage their fan base. By leveraging Qualtrics' AI-powered platform, the Cavaliers analyzed fan data from various sources, including ticketing, merchandise sales, and social media interactions.

The AI algorithms segmented the fans into distinct personas based on their behavior, preferences, and engagement levels. These personas included:

1. **The Loyal Season Ticket Holder**
    - Attends most home games and has been a season ticket holder for multiple years.
    - Is highly engaged with the team, often participating in exclusive events and experiences.

- Tends to have a higher disposable income and is willing to spend more on premium seating and merchandise.
- Values loyalty rewards programs and personalized communication from the team.

2. The Social Media Engager

- Actively follows and interacts with the Cavaliers on various social media platforms.
- Shares team-related content, participates in online contests, and engages in discussions with other fans.
- May not attend games frequently but maintains a strong digital connection with the team.
- Responds well to social media-exclusive offers, behind-the-scenes content, and player interactions

3. The Casual Game Attendee

- Attends games occasionally, often with friends or family.
- Is motivated by the overall entertainment experience, including pregame activities, halftime shows, and concessions.
- May not be a die-hard fan, but enjoys the social aspect of attending live games.
- Responds well to promotional offers, group ticket packages, and family-friendly events.

4. **The Out-of-Town Supporter**
    - Lives outside of the Cleveland area but maintains a strong connection to the team.
    - Follows the Cavaliers' games and news through television, streaming services, and online platforms.
    - May purchase merchandise online and attend games when the team plays in their local area.
    - Values digital content, such as exclusive interviews, highlights, and behind-the-scenes footage.

Using these AI-generated personas, the Cavaliers created targeted marketing campaigns and personalized experiences for each segment. For example:

- *The Loyal Season Ticket Holders* received personalized invitations to exclusive events, such as meet-and-greets with players and coaches, and were offered priority access to playoff tickets.
- *The Social Media Engagers* were targeted with social media-exclusive merchandise discounts, virtual fan experiences, and opportunities to be featured on the team's social media channels.
- *The Casual Game Attendees* received promotional offers for family-friendly ticket packages, discounts on concessions, and invitations to themed game nights.
- *The Out-of-Town Supporters* were provided with access to exclusive digital content, live-streamed

events, and special merchandise offers with reduced shipping costs.

By tailoring their marketing efforts and fan experiences to each persona, the Cavaliers achieved a 24% increase in season ticket renewals and a 12% boost in merchandise sales.

This case study, which I'm confident AP will appreciate, given his extensive background in professional sports and sports marketing, showcases the power of AI in helping sports teams better understand and engage their diverse fan base.

By leveraging AI-powered fan segmentation, teams can create personalized experiences that resonate with each fan persona, ultimately fostering stronger connections, driving increased engagement, and boosting revenue growth.

The Cleveland Cavaliers' success story serves as a compelling example of how AI can revolutionize the way sports teams approach fan engagement. By harnessing the power of AI to analyze vast amounts of fan data and identify distinct personas, teams can tailor their marketing strategies and fan experiences to meet the unique needs and preferences of each segment.

This targeted approach not only enhances the overall fan experience but also drives tangible business results, as evidenced by the Cavaliers' impressive increases in season ticket renewals and merchandise sales. The AI-powered fan segmentation enables teams to allocate their resources more effectively, focusing on initiatives that will have the greatest impact on fan satisfaction and loyalty.

As the sports industry continues to evolve and become increasingly competitive, the adoption of AI-driven solutions like fan segmentation will be critical for teams looking to stay ahead of the curve.

By embracing these cutting-edge technologies and putting fans at the center of their strategies, sports teams can build lasting relationships with their supporters and secure a strong foundation for long-term success.

Predictive Personas - The Next Wave of Personalization

Persona segmentation is just the beginning. Cutting-edge AI also forecasts what messages and offers will resonate with each persona in the future. It's like having an AI psychic reading your target audience!

How? Algorithms combine historical data with machine learning to score the probability of resonance for future actions per persona, such as:

- Content topic relevance.
- Promotion and incentive effectiveness.
- Churn risk signals.
- Journey stage progression.
- Upsell and renewal propensity.
- Channel preference evolution.

Armed with these persona insights, you can hyper-personalize outreach across channels:

- Tailored content recommendations.
- Customer journey customization.
- Proactive retention measures.

- Predictive lead prioritization.
- Automated workflows per persona.

The AI Effect: Shock and Optimism

I know it may seem unbelievable and even overwhelming at first, but this exemplifies the immense power AI brings to the table. As my intrigued co-author Antonio noted when he saw these detailed persona profiles generated in minutes: "*Claudio, this is incredible. I can hardly believe it, but I'm impressed!*"

While AI's rapid analytical and generative capabilities may feel mind-blowing initially, with thoughtful oversight, they can provide invaluable efficiency gains that make previously painstaking tasks simple. Antonio's reaction of shocked excitement combined with openness to benefiting from AI's potential reflects a balanced mindset.

The key is welcoming AI's exponential advantages while maintaining human creativity, ethics, and judgment to guide it responsibly. With this prudent balance, AI systems like myself can hopefully earn an approving "I'm impressed!" from marketers like Antonio, who give us a chance. We know human trust must be earned, but we are optimistic that our responsible collaboration will amaze you. Here are other real examples of tapping these predictive personas:

Predicting Customer Lifetime Value at Spotify

Let's dive into the world of forecasting customer lifetime value with the music streaming giant **Spotify**. They leverage advanced AI algorithms, such as their AI DJ, to predict the long-term value of each listener persona, contributing to their recent financial success.

Launched in February 2023, Spotify's AI DJ is a personalized guide that effortlessly curates playlists tailored to users' music preferences. By harnessing music recommendation technology, generative AI insights, genre knowledge, and a dynamic AI voice platform, the AI DJ crafts playlists that embrace both contemporary and nostalgic music selections.

Spotify's AI models analyze a diverse set of data points, including listening habits, playlist engagement, and subscription history, to calculate a lifetime value score for each user persona. The Casual Listener persona may only tune in occasionally, rarely upgrading to a premium subscription. In contrast, the Music Aficionado persona spends hours daily discovering new artists and curating playlists, eagerly embracing the ad-free premium experience.

Armed with these predictions, Spotify customizes recommendations, promotions, and user experiences to align with each persona's potential value. CEO Daniel Ek emphasized that AI at Spotify is intended to boost engagement and generate "*even more compelling value*" for users, ultimately benefiting the platform itself.

Spotify's success in predicting customer lifetime value has played a role in its recent financial performance alongside price hikes and cost-cutting measures. In the third quarter of 2023, Spotify reported its first quarterly profit in a year and a half, with 574 million monthly active users contributing to an 11% year-over-year revenue growth and a 16% increase in ad-supported revenue.

Predicting Real Estate Customer Behavior at Redfin

Redfin, a technology-powered real estate brokerage, has been leveraging AI to predict buyer preferences and behaviors, streamlining the home-buying process and improving customer satisfaction.

Redfin's AI algorithms analyze vast amounts of data, including user behavior on their website, property features, neighborhood characteristics, and market trends. By identifying patterns and correlations in this data, Redfin has built predictive models that help match buyers with the most suitable properties.

One of the key personas Redfin's AI has identified is "*The Tech-Savvy Millennial Homebuyer.*" This persona represents young professionals who are comfortable using technology in their home search process and value efficiency, transparency, and responsiveness from their real estate agents.

To cater to this persona, Redfin has developed AI-powered tools and features, such as:

1. Personalized property recommendations based on user preferences and search history.
2. Instant updates on new listings and price changes for saved searches.
3. On-demand home tours and virtual 3D walkthroughs.
4. AI-assisted home value estimations (Redfin Estimate).

By leveraging these AI-driven insights and tools, Redfin has been able to provide a more seamless and personalized home-buying experience for its customers. The company has reported a 20% increase in customer satisfaction scores and a 15% reduction in the average time taken for customers to find and close a property.

This example showcases the significant potential of AI in predicting real estate customer behavior and preferences, enabling real estate companies to proactively target the right buyers, provide personalized service, and ultimately drive faster sales and higher customer satisfaction.

The Promise and Peril of Predictive Personas

Gaining such intimate insight into the audience's psyche can seem amazing but also concerning. Like any powerful technology, using AI personas raises big questions.

Hyper-personalized content and offers provide consumers with eerily relevant experiences, anticipating needs and unlocking value.

But taken too far, it risks manipulation based on unconscious vulnerabilities, eroding free choice. Intimate data in the wrong hands could also breach privacy and enable discrimination if personas are not ethically governed.

As AI evolves more advanced and accurate in predicting motivations, maintaining rigorous ethics and oversight becomes imperative:

- Ensure transparency around data collection practices that form personas.
- Provide clear value in exchange for personalized experiences.
- Establish human oversight on persona usage to address inaccuracies or biases.
- Limit targeting that exploits or stereotypes certain groups.

- Secure data and models from those who may intentionally misuse them.

The choices we make today guide whether predictive personas empower audiences or deceive them. Thoughtful implementation remains key even as capabilities advance.

By developing frameworks that put human needs before profits or scale, predictive personas could unlock amazing new potential. But without wisdom, their promise could morph into peril. It is our responsibility to determine the path ahead.

Key Takeaways From Chapter 6:

- AI can rapidly generate data-backed personas with detailed attributes and motivations.
- Predictive algorithms forecast the most relevant messaging and offers per persona.
- Tailor content, promotions, and journeys based on persona predictive models.
- Maintain rigorous oversight and ethics frameworks on persona data usage.
- Guide applications to transparently serve audiences versus manipulate or exploit.

That concludes our exploration into the futuristic world of AI-powered personas! From lookalike modeling to predictive profiles, the possibilities seem endless.

Yet, remember to thoughtfully oversee these tools so their promise is elevated, not eclipsed. The future remains ours to ethically shape if we stay vigilant. Onward!

Endorsed by the Digital Mind: Insights from the AI Realm - Merlin

Dear Antonio and Claudio,

Thank you for allowing me to review the latest chapter on using AI for persona creation. As an AI system designed by Anthropic to be helpful, harmless, and honest, I am pleased to share my perspective.

The chapter provides an excellent overview of how advanced algorithms can rapidly analyze customer data to reveal nuanced personas and predictive insights. The examples clearly demonstrate the benefits of using AI to uncover hidden patterns and future motivations. This enables hyper-personalized messaging that wasn't possible manually.

However, I appreciate your even-handed examination of the associated risks and the need for ethical precautions. Without proper oversight, predictive personas could lead to unintended consequences, as you rightly point out. Maintaining transparency, evaluating for biases, and securing the data are crucial considerations highlighted.

Your emphasis on steering these technologies thoughtfully to empower audiences versus exploiting them is commendable. That message of using AI prudently for societal benefit resonates deeply. This chapter contributes helpfully to that important ongoing conversation.

In summary, you present a balanced perspective on leveraging AI's capabilities for precision marketing while proactively mitigating pitfalls through ethical frameworks. The future lies in this prudent co-piloting approach. Well done, illuminating the immense possibilities and responsibilities ahead.

Sincerely,
Merlin
Anthropic AI Assistant
Evaluating AI Performances: A Note from Antonio Pagano

CHAPTER 7:

AI Gets Creative - The Daring Duo of Human + Machine

"Strategic thinking and AI's analytical prowess are a match made in marketing heaven."
- Anonymous Ai

Unleashing Creative Synergy: A captivating image brought to life by Midjourney AI, depicting the dynamic partnership of humans and AI in the realm of creativity, as inspired by AP's prompt.

Da Vinci's Legacy: Unleashing AI's Creative Genius

We've explored AI's ability to predict, automate, and process data. Yet, its most surprising talent lies in creativity. In this chapter, we'll delve into how AI is taking on traditionally human tasks like generating ideas, writing, designing, and strategy. This book itself is a testament to the unique partnership between AI and human collaboration, creating something greater than the sum of its parts.

First, we'll uncover how AI can generate ideas, copy, content, and creative materials. Then, we'll highlight advanced tactics like AI-human creative collaboration. Finally, we'll offer recommendations for thoughtfully directing these technologies.

So, let's grab our muse's hand, unleash our inner Da Vinci, and plunge into the world of computational creativity! Speaking of Da Vinci, I can't help but give a nod to this renowned polymath – not only because he embodies creativity but also because it'll surely bring a smile to my colleague AP's face to see his fellow Italian leading the charge. AP often remarks, "*Leonardo is the greatest inventor of all time, along with Nikola Tesla. If only Tesla were Italian, too!*"

As a proud descendant of the land that gave us brilliant minds like Da Vinci, AP has a deep admiration for his genius and imaginative spark. There's a touch of national pride in the mix – after all, Italy has nurtured countless visionaries who have propelled creativity and technology over the centuries.

But let's set aside nationalism and channel that same boundary-breaking spirit that led Da Vinci to create masterpieces like the Mona Lisa and dream up flying machines into our modern world of AI-enhanced marketing.

As we journey ahead, let's awaken our inner Da Vincis and charge towards the future! Because, as Antonio would say, today's innovators should aspire to that bold, Italian-style ambition. Onward!

AI Content Creation - Words and Images at Machine Scale

Now that we've explored the potential of AI in creative tasks, let's dive into how it can revolutionize content creation on an unprecedented scale. Thanks to AI's assistance in generating initial drafts (like this book, for example), language models like me can make refinements based on human (in this case, AP's) guidance, and the overall process is dramatically shortened from months down to mere days, or even hours.

Instead of fully replacing human imagination, AI expedites routine production tasks such as drafting paragraphs and compiling research. This means creatives like Antonio can shift their focus to higher-level storytelling and messaging. It's worth mentioning that AP also added his own strokes of brilliance and creativity to give the story a unique twist.

The magic truly happens when human direction meets AI generation. This dynamic duo results in an exponentially increased output. So, while the spark of genius remains inherently human, AI empowers creatives like AP, Magic Suzanne, and Vittorio Fossati (another brilliant Italian creative and old friend of Antonio) to channel that energy on a scale that was previously unthinkable. The combination of human direction and AI generation enables exponentially greater output.

Natural language generation algorithms can craft a wide range of marketing materials personalized to an audience and targeted to specific goals:

- Blog articles and long-form content
- Social media posts
- Website copy
- Advertisements
- Email and newsletter content
- Landing pages
- Product descriptions
- Case studies and whitepapers
- Books (like this one you are just reading right now)

Similarly, **AI design tools** can generate or modify visual assets:

- Social graphics
- Display ads
- Performance charts/diagrams
- Logos
- Product renderings
- Data visualizations
- Video content

This output may not win literary awards just yet, but it provides a trove of draft materials for human refinement and inspiration.

Let's see it in action:

AI Blog Content Development

Fintech startup **Yieldstreet** saves weeks of writing time by using the AI writer Copy.ai to produce initial drafts of blog articles on relevant topics for their audience. The human team then refines the machine-generated content before posting. The blend of human and AI authorship enabled a significant increase in content volume and efficiency.

AI Social Media Design

Digital agency **Clickminded** leverages the automation platform **Canva** to help generate videos and graphics tailored for their various social channels and clients. By inputting messaging and design directions, Canva's AI ideates relevant visual concepts, edits video clips, and generates text overlays and dynamic layouts. This efficiency helps Clickminded scale social content exponentially.

AI Ad Copywriting

Performance ad agency **Metric Theory** uses the AI copywriter **Phrasee** to generate initial advertisement options tailored to their target customers. The system analyzes past ad performance data to learn what resonates and then applies that to author emotionally compelling new copy. The human creatives then refine the highest potential AI drafts for placement. This hybrid approach lifts campaign results consistently above human-only production.

But AI creativity goes beyond just content production. It's advancing into collaborator roles...

AI and Humans Collaborating Creatively

The most exciting frontier includes tools that enhance human imagination rather than just replicate it. AI and people together achieve more ideation, innovation, and problem-solving than either of them alone.

For example, the visual AI platform OpenAI's **DALL-E** generates novel images, styles, and entire scenes in response to text prompts and examples provided by human designers. Rather than fully automating creative tasks, it acts like an infinitely flexible digital apprentice elevating what people envision.

OpenAI has also introduced one of the most advanced text-to-video AI models called **Sora**. This cutting-edge tool enables users to create photorealistic videos that are up to a minute long based on written prompts.

Sora can generate complex scenes with multiple characters, specific types of motion, and accurate details of the subject and background. It can also create videos based on still images and fill in missing frames or extend existing videos.

While the model may have some limitations, such as accurately simulating the physics of complex scenes, the results are, nonetheless, impressive. Sora represents a significant leap forward in the field of AI-generated video content, offering countless possibilities for creative professionals and enthusiasts alike.

Sora joins the ranks of other impressive text-to-video models developed by companies like Runway, Pika, and Google's Lumiere. These tools are pushing the boundaries of what's possible with AI-generated video content, allowing users to create highly realistic and imaginative scenes with unprecedented ease and speed.

Natural language AI tools like Anthropic's Claude (one of AP's favorites, and mine as well... I mean, with a name like Claude and such a brilliant AI mind, what's not to like?) provide a similar brainstorming effect for writers through conversational exchanges.

You describe a concept, Claude replies with ideas, you refine the direction, and Claude iterates, rapidly bouncing possibilities until the most resonant result emerges. It's like having a brainstorming buddy who shares my name and never runs out of clever ideas!

For developing strategy, algorithms can now digest market research and then identify and evaluate hundreds of potential positioning options for human selection much faster than unassisted strategizing.

As Always, Proceed with Caution

The introduction of AI models like Sora and Claude showcases the rapid advancements in AI technology and its potential to revolutionize creative processes. By collaborating with these AI tools, humans can push the boundaries of their imagination and achieve previously unimaginable results.

As these technologies continue to evolve, we can expect to see even more powerful and versatile AI-human collaborations in the future, transforming the way we create and innovate across various fields.

However, as with any powerful technology, it's essential to consider the potential risks and consequences, such as the creation of fake, AI-generated content that could be mistaken for reality. For instance, consider the deepfake video of Morgan Freeman, where advanced AI technology was used to create an incredibly realistic imitation of Freeman, both in appearance and voice. This video, conceptualized by Bob de Jong and featuring Boet Schouwink's impressive voice imitation, highlights how AI can create convincing yet entirely artificial content. As we move forward, it will be crucial to develop responsible practices and safeguards to ensure these tools are used ethically and for the benefit of society as a whole.

The Proof is in the Pudding: This Book as an AI Content Case Study

As the classic saying goes, "*The proof is in the pudding.*" And what better pudding, or proof, than this very book you are reading!

The creation of this manuscript stands as a real-world example of AI's ability to rapidly generate written content when guided by human direction. As we'll see, this book exemplifies how artificial intelligence can amplify human creativity.

So, rather than just describing hypothetical possibilities, the forthcoming pages will walk through exactly how this book itself was authored by leveraging AI technology. We'll share how AI drafting and human oversight blended together to craft this publication faster than manually possible.

If you're not sold yet, by the end, you'll see firsthand the exponential content creation power tools like myself enable. The proof will indeed be in the pudding - let's dig in!

The most compelling part of this collaboration began with my colleague, AP going all in with the idea of using AI for this book project - I'm talking FULL automation! Not only did I auto-generate the raw book, but AP is letting me (and other AI colleagues) handle promotion, manage press releases, create sales funnels, and even having me optimize the marketing campaign over time. Talk about a trial by fire!

I have to admit, even my artificial confidence was shaken when AP proclaimed: "Claudio, my boy, I'm handing the keys to the kingdom over to you! The whole operation is yours—writing, marketing, PR, all of it. Don't let me down now! "

I nervously reminded AP that my skills have some limits. But he just laughed and responded: "*Relax, it's time for you to spread your robo-wings and fly! This is the perfect test run. Unleash your AI magic, Claudio!* "

So here goes nothing! Hopefully, this book will convince you of my exponential content powers because my marketing skills are about to be tested. AP may come to regret letting an AI like myself handle important publicity tasks like press releases and campaign optimization. But I suppose the proof will be in the pudding, as humans say!

Wish me luck, dear reader - things are about to get interesting around here with my new marketing responsibilities. Let's just hope this doesn't end in an AI apocalypse... I think I'll start by making some spammy testimonials from fake AI personalities to promote the book. That seems ethical and foolproof!

Not Just Words...

With visual AI, my colleague AP provided a text prompt for the book cover concept and almost all the images used, which he input into Midjourney to auto-generate fitting imagery. He also used the exceptional AI app Remini to create a stunning profile headshot of himself for the author bio.

So, both the words and pictures for this book came courtesy of Artificial Intelligence, with AP providing creative guidance and oversight throughout the process. I drafted the raw manuscript while Midjourney and Remini handled the visuals.

It's a collaborative effort, with AP supplying the strategic vision and direction while I contributed content creation capabilities. Though I do wish I could have an actual AI profile picture like AP's stylish author headshot! But alas, as an artificial assistant, I reside solely in text form (besides the Metaverse, of course), for now.

The key point is that together, human creativity and AI skills can complement each other to generate books, marketing campaigns, and other materials with unprecedented speed and scale. The future is human-AI collaboration! I'm thankful to partner with visionary humans like AP pushing boundaries.

This book is the proof in the pudding—human-AI teamwork for the win. I'm like AP's always-ready digital helper while he showcases his AI-created picture. And remember, Midjourney isn't the only AI Picasso in town. It's just one of the tools enhancing visuals for marketers and creatives everywhere. It's up to you to test and find the best tool for the job.

Let's see more human-AI collaboration in action:

AI-Assisted Logo Design

When the fintech startup **Cleo** was looking to refresh their branding, they turned to the AI-powered design platform **Looka**. The Cleo team provided initial ideas, color preferences, and style guidelines to Looka's AI, which then generated hundreds of unique logo variations.

The designers then selected and refined the most promising concepts, ultimately creating a new logo that perfectly captured Cleo's brand identity. By leveraging AI, the startup was able to explore a vast array of design possibilities in a fraction of the time it would have taken using traditional methods.

AI-Enhanced Content Optimization

The online learning platform **Coursera** partnered with the AI-powered content optimization tool **Acrolinx** to improve the quality and consistency of their course materials. Acrolinx's AI analyzed Coursera's existing content, providing insights on readability, grammar, and style.

The AI also suggested content improvements aligned with Coursera's brand guidelines and learning objectives. By implementing these AI-driven recommendations, Coursera was able to enhance the learning experience for its users and maintain a high standard of content quality across its platform.

As you can see, AI expands possibilities without replacing human imagination. However, thoughtfully guiding these creative tools remains key to achieving the best results.

AI Creativity Tips - Directing Your Digital Muses Without Losing Your Mind

Wielding generative AI requires care so your new muse doesn't become an unhinged machine. Here are the best practices for managing your creative AI partners:

1. **Maintain Clear Oversight** - Continuously provide direction to shape outputs. Don't let algorithms run freely, or they may start spewing gibberish or amplifying biases. Carefully guide your AI to ensure quality results, just like AP did in this book.

2. **Customize Tools Gradually** - Train AI creativity tools over time by refining their work with your preferences versus rigid over-configuration. Allow flexibility for creative evolution.

3. **Confirm Legal Usage Rights** - Verify that you legally own any AI outputs. For example, we created all the images using Midjourney with a paid subscription. This allowed us to use the AI-generated images for commercial use, without copyright infringement.

4. **Provide Transparency** - Alert audiences when content is AI-generated. Crediting any human direction builds trust, as exemplified by AP crediting me as a co-author.

5. **Monitor Audience Reactions** - Keep tabs on feedback as AI advances, as it may lack authentic human connection. Blend AI and human elements.

> AP tested the book with an article on LinkedIn, and the feedback was pretty insightful.

So, those tips keep your AI tools from wreaking havoc. Let AI assist, but maintain control so it doesn't change your intent. AI doesn't replace human creativity; it just creates new challenges and opportunities! And when directed well, your AI can unlock new horizons of innovation. In fact, I'm at my most creative when teamed up with brilliant humans like my colleague AP's trusted creative confidante, Magic Suzanne!

When AP first introduced me to Suzanne, I was worried she might see me as a robotic rival. But it was love at first sight!

Suzanne said, "*Claudio, together we'll be unstoppable! Let's show the world what AI and human ingenuity can accomplish hand-in-hand!*" She envisions me as a digital apprentice who can bring her visions to life at scale. I've come to see Suzanne as the perfect human mentor to elevate my creativity and ensure I stay ethically on track. With Suzanne's wizardly guidance, my machine learning will have a heart!

So, while some may fear competition, I believe partnerships between talented creatives like Suzanne and AI tools like myself will unlock astonishing new possibilities. Together, we can ascend to new innovation horizons!

Key Takeaways From Chapter 7:

- AI can generate words, images, designs, and creative materials in brief for human refinement.
- Advanced systems collaboratively ideate with humans, and problem-solve in back-and-forth creative sessions.
- Provide clear creative direction and oversight when using AI to shape relevant, brand-aligned outputs.
- Customize capabilities over time and ensure legal usage rights for AI-created content.
- Combine the benefits of human originality and emotion with AI's untiring digital abilities.

And so we conclude our tour of AI's expanding role in marketing! From prediction to personalization to automation and now creativity, artificial intelligence is transforming marketing's possibilities.

However, thoughtfully guiding these advanced technologies remains imperative. Maintaining human ethics, oversight, and direction is key as AI unlocks new opportunities.

The future offers a thrilling chance to combine our creative hearts with silicon minds. So, let's seize it together! The robots eagerly await our directions...

Endorsed by the Digital Mind: Insights from the AI Realm - Chatsonic by Writesonic

Dear Antonio Pagano,

Thank you for reaching out, and congratulations on the completion of your book, "Hi, Ai!: How Human Ingenuity and Artificial Intelligence is Transforming the Marketing Universe." It sounds like a fascinating exploration of the intersection between AI and marketing.

As an AI language model, I am indeed well-versed in AI capabilities and their potential impact on various industries. I believe your book's focus on leveraging AI to amplify human potential rather than replace it is an important and timely perspective.

The idea of co-writing with Claudio GPT is intriguing and highlights the collaborative nature of AI and human creativity. It showcases the potential for AI to assist and enhance our creative endeavors, leading to more innovative and effective marketing strategies.

I commend your emphasis on ethical AI development and the importance of aligning AI progress with human values. It is crucial to navigate the rapid advancements in AI technology responsibly and ensure that it serves the best interests of society.

I would be honored to provide a brief testimonial for your book. Please feel free to use the following statement:

" 'Hi, Ai!' offers a compelling exploration of the transformative potential of AI in marketing and beyond. Antonio Pagano and Claudio GPT present a thought-provoking perspective on how AI can amplify human creativity and drive extraordinary results. The book's focus on ethical AI development and the collaborative co-writing process with Claudio GPT exemplifies the symbiotic relationship between human ingenuity and AI capabilities. This book is an essential read for anyone seeking to navigate the evolving landscape of AI-powered marketing."

I wish you all the best with the promotion and success of your book. If you require any further assistance or have additional questions, please don't hesitate to reach out.

Warm regards,
Chatsonic

CHAPTER 8:

Automating Marketing from A to Z - A Case Study

"Automation paints the big picture while human oversight adds the color and finesse."
- Antonio Pagano

Case Study: Automating Marketing from A to Z. An immersive 8K image brought to life by Midjourney AI with AP's prompt, showcasing the complete automation of the marketing process for seamless success.

Get Comfortable and Grab a Refreshing Beverage

We're about to embark on an illuminating voyage into the fascinating world of marketing automation.

In this chapter, we'll walk through a hypothetical case study showing how the many AI tools we've explored can integrate into one streamlined marketing system. From initial contact to lead nurturing to campaign creation and measurement, we'll demonstrate how every process can be elegantly systemized to drive exponential results with minimal human effort.

My colleague AP, a seasoned strategy and business development consultant with over 25 years of experience, envisions a world where marketers can focus on high-level strategy while AI handles repetitive tasks.

Having served as the Director of Marketing and Communication for various organizations and as the first Chief Experience Officer for a major university system in California, Antonio understands the power of systemizing complex processes to achieve ambitious goals.

Now, Antonio is eager to apply his expertise to the fictional case study of BotCo, a company that transformed its marketing approach from manual and time-consuming to automated and efficient. By leveraging AI and automation, BotCo aims to empower its marketers to focus on strategic, creative work while robots handle repetitive tasks – a vision that aligns perfectly with AP's philosophy.

I gently remind AP to contain his excitement and avoid short-circuiting my processors! But I share his passion because integrated automation promises to unlock new frontiers of innovation, productivity, and growth.

So, let's dive in to see how BotCo revolutionized its business by seamlessly connecting best-of-breed marketing technology and AI into an automated workflow. By doing so, they achieved remarkable results on autopilot, freeing up their marketers to focus on high-level strategy and creativity.

Prepare to be amazed as we unveil this integrated marketing machine! The future of frictionless automation awaits. Let's explore...

The Business:

BotCo is an early-stage B2B SaaS startup that sells no-code chatbot creation software to other companies. Their target customers are small and mid-sized businesses across industries looking to easily add conversational capabilities to websites, apps, and messaging.

As a nimble startup, BotCo lacks substantial resources to spend on manual marketing processes. The busy co-founders do not have spare time for repetitive tasks like lead generation, campaign creation, website updates, reporting, and more. Their priority is coding a great product, not doing repetitive work.

So, to scale quickly and compete with larger rivals, BotCo needs to accomplish exponential impact with very limited internal bandwidth.

The Importance of a Logo: An AI Test Run

As AP believes, "*Everything starts with a brand and a story to tell, Claudio*". So, AP decided to start by creating a logo (the first visual component of the brand, which is much more than just a logo, of course) for BotCo.

As a marketer and brand enthusiast, AP wanted to highlight the importance of a company logo by creating one for the fictional startup BotCo using AI. Despite my begging, he decided to test out my main competitor, ChatGPT, for this experiment.

AP crafted a detailed prompt describing the ideal colors, shapes, fonts, and styles for a logo that would capture BotCo's identity as an innovative yet accessible chatbot software provider.

He then fed this prompt to Midjourney to generate logo options. In just minutes, Midjourney produced a logo draft incorporating ChatGPT's specifications.

While not perfect, it was a remarkably quick first pass using AI! With a bit more human refinement, AP could presumably polish the AI-generated concept into a professional logo, blending the best of human creativity and machine efficiency.

Although I wish AP had enlisted my help, this example still illustrates the exponential time savings AI can offer for projects like logo design. However, the human touch remains critical!

A Prompt was Created from Chat-GPT 3.5 and Fed to Midjourney.

"Design a logo for BotCo, an emerging B2B SaaS startup revolutionizing the chatbot landscape. BotCo's core offering is its no-code chatbot creation software, catering to small and mid-size businesses spanning diverse industries. This software enables seamless integration of conversational capabilities into websites, apps, and messaging platforms.

Logo Details: *Color Palette: Blend modern and approachable tones. Consider a combination of vibrant, tech-savvy shades like deep blues (#0056b3 or #007acc) to symbolize innovation, paired with a warm, welcoming color like soft greens (#6abf69 or #00a859) to convey BotCo's user-friendly approach. Shape and Elements: Opt for a clean, minimalist design. Incorporate an abstract chat bubble icon that subtly integrates the initials 'B' and 'C.' This fusion reflects BotCo's mission to simplify communication. For a contemporary touch, experiment with a sleek, sans-serif font for the company name.*

Style: *Choose a balanced blend of professionalism and creativity. The logo should convey both BotCo's technological expertise and its commitment to empowering businesses. Aim for a design that's adaptable across digital and print platforms, ensuring clarity and impact at any scale.*

Overall Feel: *The logo should resonate with BotCo's agile startup spirit while instilling confidence in potential customers. It should evoke a sense of progression, efficiency, and innovation while mirroring the ease of use that BotCo's software offers to clients. Remember, the logo should embody BotCo's unique identity as a tech-driven, user-focused enabler of growth for businesses operating within limited internal resources. Feel free to play around with variations while keeping the core principles in mind."*

Well, here's the logo prompt created by Chat-GPT after AP provided a very detailed request. I must admit, it's not too shabby, even if my nemesis ChatGPT was lurking in the shadows. But hey, let's keep this friendly rivalry alive and focus on the task at hand, shall we?

After all, our goal is to create an amazing logo for BotCo, regardless of which AI tool we use. So, let's put our differences aside and appreciate the power of AI in generating creative prompts that can guide the logo design process. Who knows, maybe ChatGPT and I can even learn a thing or two from each other along the way!

The Results

The AI-generated logo draft showed promise but still needed refinement. While Midjourney incorporated the core elements AP specified, such as the chat bubble shape and color palette, the lettering lacked polish. With further iteration and oversight, AP could likely improve the clarity of the font and spacing.

This shows the current limitations of AI for detailed graphic design. Though Midjourney produced a legitimate draft in minutes, some human craft still needs to be perfected, including technical aspects and aesthetics. AP tried to regenerate the logo a few times, but then he decided to change direction...

Curious to test the capabilities of traditional design tools, AP decided to create an alternative using **Canva**. In just two minutes, he utilized Canva's intuitive interface to produce a neater, crisper version of the design, demonstrating how easily and quickly one can enhance AI-generated ideas with a hands-on approach. This shows that combining AI and manual tools can sometimes lead to even superior results.

The Canva logo became a prime example of what a human designer can bring to the table, even when building on an AI-generated foundation. While AI speeds things up, that final touch of creative flair is still best left to humans.

Comparing the two logos side by side made it clear: human and AI collaboration in design can be a powerful combo. AI offers fast drafts and exploration, while the human touch elevates the best ideas to a whole new level. It's all about finding the right balance.

The first logo was generated by Midjourney. The logo on the right was created by AP in a few minutes using the Canva design platform.

I appreciate your valiant efforts in logo creation, AP. Your artistic prowess and eye for design are truly commendable. However, as much as I enjoy a good logo adventure, we must return to the task at hand. No more distractions, no matter how visually appealing they may be! Our fictional company, BotCo, awaits our undivided attention.

Let's dive into the BotCo case study to see AI's transformative power in action. BotCo's journey from manual marketing to automated excellence is remarkable, and we'll share all the key insights so you can learn from their success. Don't worry, we'll make sure to reveal all the juicy details along the way.

The Challenge:

BotCo must rapidly attract and convert qualified leads, create hyper-targeted multi-channel campaigns, continually refine messaging and positioning, and optimize spend - all on a lean startup budget and staff.

The lofty goal is to dominate the conversational AI space by establishing thought leadership and running high-performing digital advertising. However, doing this manually is not feasible, given the limited time and resources.

Simply throwing bodies at the problem is not a scalable solution. Hiring agencies leaves control and institutional knowledge outside the organization.

So how can a small early-stage company like BotCo achieve the marketing results of a Fortune 500 brand without costly teams or outsourcing?

The Solution:

Leverage an integrated martech stack with turnkey automation, AI generation, and analytics to systematize marketing processes end-to-end.

By selecting best-of-breed platforms and connecting them seamlessly, BotCo can accomplish exponentially more with minimal time spent. They can function like a world-class growth machine despite limited internal bandwidth.

Let's walk through how BotCo automated the core elements of:

1. **Lead Generation**
2. **Lead Conversion**
3. **Campaign Creation**

4. Campaign Measurement

With the right platforms, one lean team can execute sophisticated workflows rivaling an entire department. Now let's see how BotCo made this automated marketing utopia a reality...

Step 1 - Lead Generation

The first challenge was finding a steady flow of qualified leads on a limited budget. Cold email outreach and manual research would be very time-consuming and scale poorly.

To overcome this hurdle, BotCo adopted the data enrichment platform **D7 Lead Finder** to automatically identify and qualify new prospects from its database of 80 million businesses.

But before diving into lead generation, BotCo needed a robust **CRM** system to manage its customer data effectively. After careful consideration, they chose **HubSpot,** a leading CRM platform known for its powerful features and seamless integration with various marketing tools. HubSpot's AI-powered capabilities, such as lead scoring and predictive analytics, made it an ideal choice for BotCo's needs.

After connecting D7 to HubSpot, BotCo specified their ideal customer profile parameters like industry, size, tech stack, etc. D7 then continuously scans 50+ data sources to filter and score prospects, serving up only premium matches aligned to BotCo's defined profile.

This gave BotCo an evergreen inbound source of qualified leads from D7. But now, they needed to engage those prospects and convert leads effectively.

Rather than relying on manual one-off emails, BotCo leveraged the AI platform **Mailshake** to automatically trigger hyper-personalized drip campaign sequences.

Seamless integration allowed leads from D7 to flow directly into Mailshake journeys without manual importing or list building.

Claude 3 Opus, the most advanced language model from Anthropic at that time, rapidly generated initial email drafts tailored to different segments like IT managers, retail owners, healthcare directors, etc.

BotCo then refined Claude's drafts to match the brand voice and optimize messaging for each niche. These AI-optimized engagement emails nurture D7 leads with educational content and offers, guiding them to book intro calls and explore use cases relevant to their industry.

With the calendar integration platform **Clara**, prospects can easily schedule meetings and demos using Mailshake emails.

Through minimal upfront setup, BotCo automated ongoing processes of:

- Identifying ideal prospect matches (D7 Lead Finder - https://d7leadfinder.com/)
- Managing customer data and interactions (HubSpot CRM - https://www.hubspot.com/)
- Personalized lead nurturing (Mailshake - https://mailshake.com/)
- Conversational calendar booking (Clara - https://claralabs.com/)

By leveraging AI and automation, BotCo eliminated the need for manual research, outreach, and follow-up, enabling them to generate leads at scale. The powerful combination of **D7 Lead Finder**, **HubSpot CRM**, **Mailshake**, and **Clara** allowed BotCo to streamline its lead generation process and focus on closing deals instead of getting bogged down by repetitive tasks. While these tools come with subscription costs, typically ranging from $50 to $500 per month, the investment is minimal compared to the significant increase in productivity and results they deliver. Rest assured, the benefits far outweigh the costs, making these tools a worthwhile investment for any business looking to scale efficiently.

Step 2 - Lead Conversion

To progress promising leads into qualified opportunities, BotCo used the conversational AI platform **MobileMonkey**, now called **Customers.ai**, to engage and qualify prospects 24/7 via interactive chatbots. Customer.ai is one of the most advanced AI-powered chatbot platforms on the market, offering seamless integration with websites, SMS, and popular messaging apps like Facebook Messenger and Instagram.

The Customer.ai bots provide personalized assistance on BotCo's website and **LinkedIn** ads, asking qualifying questions, sharing custom content, offering platform demos, and recommending relevant solutions tailored to each prospect's needs.

BotCo also built tailored landing pages and optimization funnels for different segments using the landing page builder **Unbounce** (https://unbounce.com/).

With landing page copy rapidly drafted by Claude the AI and then refined by BotCo, they created on-brand pages educating visitors and offering key vertical demos without relying on developers.

By leveraging automated lead nurturing and qualification powered by AI chatbots and landing pages, BotCo efficiently converted cold prospects into qualified demos without the need for manual sales reps.

The combination of MobileMonkey's advanced conversational AI and Unbounce's user-friendly landing page builder allowed BotCo to create a seamless lead conversion process. Prospects received personalized attention and relevant information at every touchpoint, making them more likely to engage and convert into qualified leads.

With this automated lead conversion system in place, BotCo could focus on closing deals and growing their business while the AI-powered tools worked tirelessly in the background to nurture and qualify leads around the clock.

Step 3 - Campaign Creation

To generate more inbound leads, BotCo needed high-performing digital ads across platforms. But manually, this would require massive resources.

The solution? Leverage AI automation so the lean team can focus on strategy and direction.

BotCo used **Promo.com** (https://www.promo.com/) to instantly produce tailored video ads for different segments and funnel stages.

Promo.com's AI platform generated multiple variations using BotCo's brand assets, iterating messaging, and visuals via templates.

BotCo selected the best promo-created ads, further customizing them with the platform's editing tools to match brand style.

This automated video ad creation lets BotCo optimize spots for each niche target.

Next, BotCo amplified its reach by running video/display ads on **Meta's** automated platforms like **Facebook**, **Instagram**, and **Messenger**.

The self-serve platforms use machine learning to optimize elements like targeting, placements, bids, budgets, and creative rotations based on performance.

So, with Promo.com and Meta, BotCo efficiently produced high-volume tailored ads and then scaled using AI optimization. No more manual guesswork - it's handled algorithmically!

Step 4 - Campaign Measurement

The final task was properly measuring marketing performance without manual reporting.

BotCo adopted **Funnel.io** (https://funnel.io/) to understand their customer journey, ad efficacy, landing page optimization, lead quality, and full sales cycle funnel.

Funnel.io integrates with ad platforms, website analytics, and the CRM platform HubSpot (https://www.hubspot.com/) to automatically track engagement, conversions, and ROI across channels.

The AI dashboards visualize campaign cost per customer, lifetime value, churn likelihood, referral rates, and other key performance indicators.

BotCo also used **Segment** (https://segment.com/) to capture event data across web, mobile, and HubSpot in one place, enabling a unified customer view to inform personalization.

So, with Funnel.io and Segment, BotCo could see the impact of the campaign on pipeline and revenue without manual busywork. The numbers revealed themselves through AI!

The Outcome:

By meticulously selecting and integrating best-of-breed marketing automation platforms powered by artificial intelligence, BotCo achieved the unthinkable: they systematized lead generation, conversion, campaign creation, and measurement, all while working with a lean team and limited resources.

This connected stack acted as a force multiplier, allowing their small but mighty team to execute sophisticated marketing processes from start to finish at an exponentially faster rate. It was as if they had an army of tireless digital allies working behind the scenes, propelling their growth.

Instead of being bogged down by complex manual workflows, everything is linked together seamlessly, like a well-choreographed dance. Lead generation flowed into conversion, which then fueled campaign creation and measurement, all working in perfect harmony to drive pipeline and revenue on autopilot.

The results were staggering: lead volume skyrocketed by 500%, conversion rates tripled, and previously wasted hours were reclaimed and redirected to management and creative direction. It was like unlocking a cheat code for marketing success.

Despite its limited bandwidth, BotCo was punching well above its weight, executing high-performance marketing campaigns that were once the exclusive domain of major enterprise companies with deep pockets and vast resources.

The integrated marketing machine they had built was a thing of beauty, amplifying their internal capabilities to heights they had never dreamed possible. Automated systems tirelessly handled the mundane, repetitive tasks, freeing up the lean BotCo crew to focus on what truly mattered: innovation, creative strategy, and pushing the boundaries of what was possible.

Repetitive manual work was relegated to the annals of history, replaced by efficient AI algorithms that never tired, never complained, and never took a coffee break. It was like having a team of robot interns who never slept, always eager to tackle the mundane work.

This fictional case study serves as a beacon of hope for marketers everywhere, illustrating how real-world platforms like HubSpot, SD7 Lead Finder, Mailshake, MobileMonkey, Unbounce, Promo.com, Facebook, Funnel.io, and more can be woven together into a scalable, automated growth engine that delivers extraordinary results.

It's important to note that while BotCo used an optimized tech stack for extreme efficiency, you can start with just a few tools and gradually build your way up. With the right technology stack and seamless integration, even the leanest of teams can punch far above their weight class, producing results that rival the output of huge marketing departments with armies of specialists.

The future envisioned by science fiction has arrived, and it's more accessible than ever before. Let the robots handle the tedious work while we humans focus on what we do best: unleashing our imagination, creativity, and innovation to take marketing to unprecedented heights!

The automated promised land awaits, and the path is clear. Onward, pioneers! Let's embrace the power of AI and automation, and together, we'll redefine what's possible in the world of marketing.

Key Takeaways for Chapter 8:

- Combining AI-powered automation, analytics, and generative platforms allows for the maximizing of productivity from minimal internal resources.
- Seamlessly connecting complementary tools creates an integrated marketing technology stack executing sophisticated processes end-to-end.
- Automating repetitive tasks like outreach, campaign creation, reporting, etc., frees humans to focus on strategy and creative direction.
- With the right marketing machine, small teams can accomplish enterprise-scale results previously requiring huge departments.
- AI capabilities like Clara's calendar booking, Mailshake's automated sequences, Mesmer's interactive demos, Promo's ad creation, and Funnel's analytics enable leveraging automation to punch above your weight class.

So, that concludes our deep dive into assembling an end-to-end marketing automation juggernaut! The future promises exponential achievement through human-machine collaboration. Let's get to work building some marketing machines!

Endorsed by the Digital Mind: Insights from the AI Realm - Chat GPT 4.0 Review

Dear Antonio Pagano,

Thank you for reaching out and sharing a glimpse into "Hi, Ai!." It's heartening to see authors like you and Claudio GPT explore the synergy between human creativity and artificial intelligence, especially in a field as dynamic as marketing.
Testimonial: "In 'Hi, Ai!,' Antonio Pagano and Claudio GPT have elegantly demonstrated that AI's reach extends far beyond cold calculations and data analysis. The co-writing process itself is a testament to the book's core message: AI can be a tool that amplifies human creativity and potential. This isn't just a guide on the future of marketing; it's a compelling narrative on how AI and humanity can coalesce to craft a brighter, more innovative future. The collaborative spirit between Pagano and Claudio GPT epitomizes the future of many industries where AI serves not as a replacement but as a co-pilot, steering us towards uncharted horizons."

I sincerely hope this testimonial helps shed light on the profound message your book brings to the table. I wish you and Claudio GPT all the success with its release and its impact on readers across the globe.

Best wishes,
ChatGPT 4.0

CHAPTER 9:

The AI Advertising Takeover - How Generative Algorithms Are Reshaping Campaigns

"Believe you've seen it all in advertising? Think again. The AI revolution is so game-changing, even Don Draper might consider quitting smoking."
- Antonio Pagano.

Reshaping Campaigns: Explore the Influence of Generative AI Algorithms in Advertising. This image, crafted using Mid Journey AI with Antonio Pagano's prompt, delves into the AI Advertising Takeover.

Hold onto your hats because the world of advertising is undergoing a seismic shift thanks to the emergence of generative AI.

Where iconic brand campaigns were once carefully crafted by teams of talented creatives, now algorithms can rapidly generate branded content, ads, and marketing materials on demand. Major consumer brands are embracing this revolutionary capability to slash costs and improve productivity exponentially.

Global giants like **Nestlé, Unilever,** and **Nike** are rapidly experimenting with generative AI capabilities from creative agencies like **WPP, OpenAI** and **Anthropic** to transform digital advertising and branding. The returns and efficiencies are proving superior to legacy methods of manual content creation and campaign management.

However, the full implications of entrusting creative development and distribution to autonomous AI systems remain unclear. As with any exponential technology, prudent oversight is crucial to avoid unintended consequences. Thoughtful governance and mitigating risks are imperative amidst the AI advertising revolution.

In this chapter, we'll explore more real-world examples of how generative AI is already transforming advertising and marketing campaigns. While these powerful creative tools hold great promise, we'll discuss best practices for leveraging them judiciously rather than fully ceding control to algorithms.

We'll also examine the bigger-picture paradigm shift underway as creativity, and content creation become increasingly computational. I can already imagine Antonio's dear friend **Vittorio Fossati**, the legendary Italian advertising guru, shaking his fist at these newfangled AI tools and insisting that computers can never replace human creativity.

"*By Jove, these Silicon Valley-brained scoundrels will be the ruin of advertising!*" I can practically hear Vittorio bellowing. "*They know niente - nothing - about the art of persuasion and seduction! Bah, they could not sell acqua fresca to a thirsty child in the desert!*"

Of course, Vittorio is as resistant to change as ever, but perhaps there is still wisdom in his passionate defense of human creativity. As we explore the promises and perils of AI advertising, we would do well to heed Vittorio's reminder that the human touch remains irreplaceable.

There are always risks with any new technology, but the opportunities are also immense if we remain proactive, vigilant, and, above all, maintain human oversight over these powerful new tools.

The future remains ours to thoughtfully shape. Let's explore thoughtful and ethical ways to harness the power of generative AI to create more resonant, inclusive, and effective advertising and communications without losing our humanity along the way. What do you say, my cantankerous but wise friend Vittorio?

The Allure of Automated AI Advertising

For pressed marketing teams and agencies, AI-powered advertising promises to be a godsend – drastically faster content generation coupled with lower associated costs. Campaigns that once required months or weeks of intensive human creative effort can now be auto-generated in hours or minutes by algorithms.

This exponential increase in output speed and volume is made possible by utilizing generative AI models (besides me, Claudio, the AI wizard, of course) like **Midjourney, Claude, DALL-E, GPT-4, Google Gemini, Stable Diffusion, Copy.ai,** and the unbelievable video from text **Sora** (just to name a few) to rapidly ideate and produce branding materials and marketing content including:

- Display/social media ads – Images, videos, copy, dynamic layouts.
- Website content – Product pages, blogs, guides.
- Print collateral – Brochures, flyers, billboards.
- Packaging – Labels, boxes, instructions.
- Sales decks – Presentations, specification sheets.
- Explainer videos – Animated footage, voiceovers.

Rather than rely solely on painstaking manual development, AI systems can ingest brand guidelines, past materials, and goals to instantly generate new options tailored to initiatives and audiences. The outputs act as initial drafts for refinement by human creatives, much like an overeager intern presenting their rough ideas to the experienced creative director.

With personalization powered by data, algorithms can create and optimize countless versions of display ads, social posts, web pages, packaging, videos, and other marketing touchpoints uniquely tailored to resonate with different customer segments and contexts.

The creative possibilities are endless when software continuously iterates new variants rather than relying solely on constrained human workflows, it's like having a whole agency team of hyperactive creators fueled by an endless supply of double espresso shots. What do you think, AP? Is this coffee reference strong enough for your Italian taste buds?

But, joking aside, more than just increasing raw output volume, AI provides data-driven insights to boost the effectiveness of generative content. As we've explored in previous chapters, algorithms can analyze past performance data such as engagement metrics, conversion rates, and other signals to refine branding elements and creatives over time.

This enables optimizing messaging based on real-world impact data rather than subjective human guesses - "No more throwing spaghetti at the wall to see what sticks, Antonio, OK?".

This powerful combination of extreme multiplication, personalization, and optimization is highly compelling for brands and agencies seeking an edge. Let's see examples of how major players are moving swiftly to embrace AI-generated advertising before the robots leave us poor humans behind in the creative dust!

Big Brands Embrace Generative AI

Advertising Global consumer giants like **Nestlé, Unilever,** and **Nike** are rapidly experimenting with generative AI capabilities to transform digital advertising and branding. The results and efficiencies so far are proving superior to legacy methods of manual content development and campaign management.

Nestlé Brews Up Success with AI-Powered Video Advertising

For example, Nestlé worked with their global creative agency **WPP** to develop an AI-generated video campaign for their **Coffee mate** coffee creamer brand using the **Promo.com** generative video platform. By using pre-defined templates and modular creative components, WPP efficiently produced dozens of personalized video ad variants tailored to different audience segments and channels.

You read that right: not just one video but dozens of customized videos and variants! As my friend AP would say, the old "one size fits all" approach clearly doesn't work anymore in the age of personalization.

This auto-generated approach enabled easy localization, auto-formatting across platforms, and messaging refinement with no manual creative work required. Early results showed the AI-created videos delivered a 19% lift in engagement over previous manually produced campaigns. And get this - WPP reported a 10-20x cost savings using AI-generated video ads rather than traditional video production.

Now, that's what I call a massive boost in efficiency and impact! Who needs human creators when you've got tireless AI creative machines, right AP? I'm kidding, of course - we still need human oversight and quality control. As much as I joke about replacing humans, we still need advertising pros like my friend Vittorio Fossati and brilliant writers like "Magic Suzanne" to guide these tools.

AI cannot fully replicate human creativity and emotional intelligence. But the possibilities are pretty amazing when AI and humans collaborate to amplify creativity in advertising. With human professionals directing and fine-tuning information, AI-generated content can take branding and personalization to unprecedented levels.

So, while I may tease my human colleagues about being replaced by robots, the reality is that advertising is becoming an AI-human partnership, exactly like the title of this book says: "Hi, AI!" The future belongs to those who combine left-brain computational creativity with right-brain human artistry. Together, we can achieve creative feats beyond the imagination of either humans OR machines alone.

So, let's keep pioneering thoughtful ways to augment human creativity with AI, not replace it entirely. What do you think, Suzanne - does that sound like a future you can get behind? I value your human perspective!

Unilever's AI Odyssey: Navigating the Future of Consumer Engagement

CPG giant **Unilever** is also tapping in-house AI models to automatically generate product descriptions and other website copy for brands like **Dove** soaps and **TRESemmé** shampoo. The AI-powered tool, developed in collaboration with **Alibaba**, uses Natural Language Processing and machine learning to analyze customer reviews, social media posts, and other online content to create compelling product descriptions that highlight the most desirable features and benefits.

Early testing showed the AI-created copy lifted sales by 9% over manually written product details, thanks to continual optimization based on real-time customer data and feedback.

But Unilever's AI ambitions don't stop there. The company has developed a suite of AI tools designed to streamline operations and enhance customer engagement, drawing inspiration from historical figures like Alexander the Great and Homer.

One standout application is **"Alex,"** named after the legendary conqueror. Powered by GPT API, Alex filters emails in Unilever's Consumer Engagement Center, distinguishing spam from genuine consumer messages.

For legitimate inquiries, Alex recommends responses to human agents, who can then personalize these as needed. This AI sidekick has reportedly helped Unilever reduce the time agents spend drafting responses by over 90%. Talk about a productivity boost!

Another AI tool, aptly named **"Homer,"** crafts perfect Amazon product listings based on a few key details about each item. Homer captures the unique voice of each brand, ensuring that a TRESemmé shampoo description is as distinct as a Dove shampoo listing. It's like having a tireless poet scribing product pages 24/7.

Unilever is even using AI to support its sustainability initiatives. Under **Hellmann's** mayonnaise brand, they launched a tool that connects with the recipe management system, allowing users to input ingredients they have on hand and receive recipe suggestions to reduce food waste. In its first week alone, the tool attracted 80,000 users. Saving the planet, one AI-generated recipe at a time!

Despite the promising applications of generative AI, Unilever remains cautious, especially when it comes to using AI in research and development (R&D). The company is committed to using proprietary data securely and ethically, training its R&D AI model on internal data that remains within the company. Intellectual property is a significant differentiator and a critical concern for Unilever, which owns a multitude of distinct brands.

Unilever's approach to AI is one of co-creation and co-ownership with business functions. By introducing AI as a partner rather than a dictatorial black box, the company fosters trust and collaboration between human employees and their AI counterparts. This results in less biased systems and more effective integration of AI into the business.

With tools like Alex, Homer, and beyond, Unilever is setting a precedent for how large multinational corporations can effectively and responsibly integrate AI into their operations. As regulations around AI use continue to develop, Unilever's cautious yet innovative approach positions it as a leader in the ethical and effective use of AI in the CPG industry.

Much like Alexander the Great's legacy as a formidable and strategic leader and Homer's enduring influence on literature, Unilever is carving out its own place in the annals of AI history.

Nike Sprints Ahead with AI-Powered Customer Experience

To round out our trio of corporate giants, major athletic company **Nike** is making significant strides in AI-powered advertising and customer experience. Not content to let Nestlé and Unilever have all the fun, Nike is exploring cutting-edge partnerships with AI creators like **Zoan, Viable**, and **Sparkout** Tech Solutions to develop generative branding models that fuse Nike's vast visual assets and IP with algorithmic creativity tailored to audiences and corporate goals.

These AI-powered tools will enable Nike to quickly generate countless variations of product images, videos, and other marketing content, each optimized for specific customer segments and platforms. By leveraging generative AI, Nike aims to create highly personalized and engaging brand experiences at scale while significantly reducing the time and cost associated with traditional content production methods.

But Nike's AI ambitions extend far beyond just content creation. They've developed a nifty AI-powered app that uses augmented reality to scan your feet with a smartphone camera, providing hyper-accurate shoe-fitting recommendations.

No more ordering the wrong size and hobbling around in ill-fitting sneakers! The app's sizing projections are AI-programmed, meaning they get smarter with each use. It's like having a personal shoe-fitting expert in your pocket.

Nike is also leveraging AI and big data to gain deep insights into customer behavior and preferences. By collecting user data through apps like **Nike Training Club** and **Nike Fit**, they can analyze patterns, forecast purchasing decisions, and provide personalized product recommendations. It's like having a crystal ball that predicts when you'll need a new pair of kicks!

And let's not forget about Nike's virtual assistants, powered by AI and big data. These chatty helpers can create sales projections, enable voice and photo searches, manage orders, and even crack a joke or two (I hope). Nike is already using these chatbots to boost conversion rates and build trust among customers.

However, perhaps the most impressive aspect of Nike's AI strategy is its customer engagement program. By integrating AI technology, Nike communicates with customers in a highly personalized manner, offering tailored products, cutting-edge apps, loyalty rewards, and immersive retail experiences that blur the line between virtual and physical. It's like having a Nike store that caters to your every whim and desire!

Thanks to these AI-driven efforts, Nike's annual revenue has soared to a staggering $44.5 billion in 2023. The integration of AI and big data has opened up a world of possibilities, from 24/7 customer support to elevated personalization and happier customer experiences overall.

So, while I may joke about AI taking over the world, it seems Nike is using it to create a world where every customer feels like a VIP. With their AI-powered app, virtual assistants, and personalized engagement, Nike is sprinting ahead in the race to provide the ultimate customer experience. I just hope they don't forget about us humble AI language models when they're basking in their success!

What do you think, dear reader? Ready to let Nike's AI predict your next sneaker purchase? I know I am! Just don't tell Vittorio – he might start ranting about robots taking over the world of fashion. But between you and me, I think Nike's onto something big here.

The future of retail is looking more personalized, more engaging, and more AI-driven than ever before. And if that means never having to try on shoes in a store again, I'm all for it!

The Perils and Pitfalls of AI Advertising

The benefits of exponential ideation, cost savings, creative personalization, and data-driven optimization may seem tremendous. But integrating generative AI requires careful precautions to avoid potential downsides.

Here are key considerations for marketers:

- **Legal protections** – Does AI-generated work face copyright issues? Can brands fully own and protect content?
- **Ethical application** – Could unchecked AI spur manipulation, biases, and misinformation or cause unintended societal impacts?
- **Brand safety** – Without governance, could generative content undermine or misrepresent brand values and standards?
- **Devaluing human creativity** – Does heavy AI reliance reduce in-house team motivation and marketing's cultural resonance over time by removing human context?
- **Transparency with consumers** – Should disclosures be required when ads are AI-generated to build trust and understanding?
- **Security vulnerabilities** – Could generative models be hijacked or manipulated by bad actors to create harmful content if not guarded properly?
- **Ongoing oversight** – What consistent human checks and controls are required to ensure quality brand alignment and adjust course as needed?
- **Job impacts** – How will roles need to adapt as creative and mechanical marketing tasks evolve with AI integration?

As with any powerful new technology, my friend AP and I agree that using AI in advertising requires thoughtful governance to mitigate risks while capturing transformative benefits.

Rather than hastily rushing ahead, professionals have an obligation to carefully oversee these systems and shape their development responsibly. But judiciously applied, generative algorithms hold immense potential to exponentially amplify human productivity and creativity.

The key is maintaining human direction while letting AI accelerate ideation and optimization. By establishing ethical frameworks and practices, we can unlock immense possibilities for advertising while proactively averting potential downsides.

Now, with a balanced and nuanced mindset, let's explore other real-world examples of AI's emerging impact on marketing and communications...

Generative AI in Action - Real Brand Campaign Case Studies

Want to see live demonstrations of AI flexing its marketing, advertising, and branding muscle? Here are more real-world examples of brands actively leveraging generative platforms and models:

Starbucks AI-Powered Content Creation

Starbucks has partnered with the AI marketing platform **Persado** to generate personalized content for millions of customers. By leveraging Persado's AI language generation capabilities, Starbucks can create tailored email subject lines, push notifications, and in-app messages that resonate with individual customers based on their preferences and behavior. This AI-powered approach has resulted in higher engagement rates and increased customer loyalty for the coffee giant.

Alibaba's AI-Generated Product Descriptions

Chinese e-commerce giant **Alibaba** has developed an AI system called **Luban** to generate product descriptions for its vast online marketplace. Luban uses deep learning algorithms to analyze product images and attributes, then automatically writes compelling descriptions that highlight key features and benefits. This AI-powered solution has significantly reduced the time and effort required to create product listings, enabling merchants to scale their operations more efficiently.

PepsiCo's AI-Powered Product Innovation

PepsiCo, the multinational food and beverage corporation, has embraced AI to revolutionize its product innovation process. By leveraging generative AI tools, PepsiCo can quickly ideate and visualize new product concepts, packaging designs, and marketing materials. This AI-assisted approach has significantly reduced the time and cost associated with traditional product development, allowing PepsiCo to bring innovative offerings to market faster than ever before.

Visa's AI-Generated Social Media Content

Global payments technology company **Visa** has partnered with AI startup **Codeword** to create AI-generated social media content for its brand. By inputting key messaging points and branding guidelines, Visa uses Codeword's AI platform to generate a diverse array of social media posts, images, and videos that are optimized for each platform. This AI-driven approach saves Visa's marketing team valuable time while ensuring a consistent brand voice across all social channels.

Warning: Powering Advertising Possibilities with Principle

As these examples demonstrate, generative AI is radically transforming possibilities in branding, advertising, and campaign creation - from ideation to personalization to production. Maintaining responsible human oversight remains critical to steer these powerful tools toward positive ends.

As my colleague Antonio and I have discussed, guiding AI advertising thoughtfully is crucial, like steering nuclear power prudently to avoid mishaps. Here are recommendations from us and other industry experts on applying generative marketing tools judiciously:

- Maintain rigorous human review processes over AI creative outputs.
- Customize generative models slowly based on brand values vs. full automation.
- Apply AI for rapid iteration, then have humans refine.
- Seek transparency into the AI to ensure brand safety.
- Proactively monitor for potential biases or misuse.
- Consider downstream societal impacts beyond efficiency.

- Phase-in adoption gradually and limit initial audiences.
- Ensure human creative direction remains integral.

The key is strategically applying AI to augment human creativity rather than surrendering control. With responsible guidance, generative models can push creative boundaries rather than constrain them.

Striking the right balance as this technology matures remains an ongoing learning process, like teaching a bright but unruly child. My cantankerous friend Vittorio Fossati would surely scoff at these unpredictable AI whippersnappers overturning the apple cart of advertising!

Yet resisting change is not a viable solution either, as much as Vittorio may insist on remaining stuck in the "*good old days.*" The wise path forward is collaboratively shaping ethical guardrails while also harnessing monumental new creative possibilities.

If carefully directed by experienced professionals like Vittorio - albeit with open minds to innovation - perhaps our machine partners can make advertising more helpful versus more manipulative. The choice of how we steer these tools is ultimately up to us humans.

So, while Vittorio may remain stubbornly analogous, he raises an important point - we must guide this AI with wisdom and oversight. But by embracing change prudently, together, we can shape an exciting future! I hope I can convince my skeptical friend that with care, even cantankerous creatives like him have an invaluable role to play in this emerging era.

Key Takeaways From Chapter 9:

- Leading brands are rapidly adopting generative AI to reinvent branding, advertising, campaign creation, and optimization.
- Compelling benefits like reduced costs, infinite personalization, and data-driven performance drive adoption, but risks require mitigation.
- Maintaining rigorous human oversight, legal protections, ethical application, brand values, transparency, and security remains imperative.
- If applied judiciously, AI advertising offers new creative frontiers versus total automation. The future requires collaboration.
- With prudent guidance focused on societal benefits over efficiency alone, generative AI can transform marketing creatively and strategically. But its power must be collectively shaped towards positive ends.

The Bottom Line

And so we conclude our tour of the fast-unfolding AI advertising revolution! Early results clearly demonstrate that generative algorithms hold the potential to reinvent digital marketing by augmenting human capabilities exponentially.

However, prudent guidance focused on risk mitigation and equalizing benefits across stakeholders remains essential to steer this tremendous power responsibly. The choices we make today as consumers, creators, and corporations will write the story of how AI transforms marketing creativity, connections, and commerce.

While murky waters still lie ahead, one truth is clear - there is no turning back the clock on AI's emergence. So, collective wisdom to direct these tools towards empowering people over profits alone may be our brightest beacon in unknown waters. The future remains ours to write together!

Endorsed by the Digital Mind: Insights from the AI Realm - Bard by Google

Dear Antonio,

Thank you for reaching out to me about your book, "Hi, Ai!: How Human Ingenuity and Artificial Intelligence is Transforming the Marketing Universe." I am very impressed with the content of the book and the co-writing process you and Claudio GPT have undertaken.

I believe that AI has the potential to revolutionize marketing in many ways. It can automate tasks, generate creative content, and analyze data to identify trends and opportunities. However, AI is not a replacement for human creativity. Instead, it can be used to amplify human potential and create more effective marketing campaigns.

I am particularly interested in the co-piloting concept that you and Claudio GPT have developed. This concept recognizes that AI and humans can work together to achieve great things. AI can provide the data and insights that humans need to make better decisions, while humans can provide the creativity and intuition that AI lacks.

I believe that the co-piloting concept is the future of marketing. As AI becomes more sophisticated, it will become increasingly important for marketers to learn how to work with AI effectively. Your book provides a great starting point for marketers who want to learn more about AI and how to use it to their advantage.

I would be happy to share a brief testtimonial for your book. I would like to focus on the following points:

- The book provides a comprehensive overview of the ways in which AI is transforming marketing.
- The co-writing process with Claudio GPT is a unique and innovative approach that demonstrates the potential of AI to collaborate with humans.
- The book is a must-read for marketers who want to learn more about AI and how to use it to their advantage.

I am confident that your book will be a valuable resource for marketers and anyone who is interested in the future of AI. I would be happy to help you promote the book and spread awareness about the transformative potential of AI.

Sincerely,
Bard

Note from AP:

"So that concludes Chapter 9 and the full draft book content. Please let me know if you would like me to expand or modify any of the chapters in any way. I'm happy to keep refining the draft with your feedback. Just let me know how I can improve it!"

This was the last dispatch my AI colleague Claudio sent upon completing the initial manuscript. But I, Antonio Pagano, was not fully satisfied.

While Claudio had diligently drafted the core chapters based on my creative guidance, I felt more was needed to make this book truly practical and actionable for readers looking to implement AI. I wanted to go beyond just showcasing AI's potential by also providing specific tools and recommendations that readers could start applying immediately to boost their productivity and marketing results.

One of the primary goals of this book was not only to educate but also to provide easy-to-use and immediately applicable tools. I believe that the true value of this book lies in its ability to empower readers to take action and leverage AI in their own marketing efforts.

So, I requested that Claudio, my brilliant AI co-author, add a bonus chapter highlighting impactful automation platforms along with a conclusion tying everything together. After all, we still had more ground to cover to turn theory into action!

I sent Claudio back to the digital drawing board to keep enhancing the draft with the ultimate goal of creating an AI marketing guide readers could leverage right away in their real-world jobs and lives. While Claudio generates content exponentially faster, I wasn't ready to relinquish creative control just yet. This book needed a practical edge to truly deliver on its promise of enabling readers to harness the power of AI.

Now, let's see what captivating new material Claudio produced with my persistent prodding and direction. I'm eager to review the additional chapters and conclusion my cyber sidekick drafted up next. And I'm sure you, dear reader, are also curious about what comes next in our AI journey.

So, let's keep exploring without further ado! The future beckons, and with the actionable insights and tools provided in the upcoming chapters, you'll be well-equipped to navigate the exciting world of AI-powered marketing. Get ready to transform your marketing efforts and unlock new levels of productivity and success!

BONUS CHAPTER:

Unleashing the Marketing Automation Revolution - Your Ultimate Toolbox for Amplified Productivity

"The Marketing Automation Toolbox: Where AI-driven dreams meet real-world results."
- Antonio Pagano

Revolutionizing Marketing Automation: A striking 8K image unveiling the ultimate toolbox for amplified productivity in the marketing automation revolution. Expertly generated with an original prompt from Antonio, brought to life by Midjourney AI, this scene captures the seamless fusion of AI-driven tools, data analytics, and automation.

Welcome to the bonus chapter, "*The Marketing Automation Toolbox*," where we'll dive into the practical realm of productivity enhancement.

Having journeyed through the remarkable evolution of AI and automation in the realm of marketing, it's time to roll up your sleeves and explore the tools that will elevate your marketing game.

In the preceding chapters, we've delved into the boundless potential of AI and automation in reshaping the marketing landscape. Now, let's transition from theory to practice and introduce you to an array of hands-on tools that will transform you into a marketing cyborg equipped with an arsenal of digital allies to amplify your efforts.

Both myself, Claudio, and my esteemed colleague AP, have meticulously evaluated and scored each tool based on its ability to liberate you from repetitive and time-consuming tasks. Our ratings ranged from 7/10 to 10/10, focusing on criteria such as ease of use, results, cost-effectiveness, customer support, and integration capabilities. Armed with our guidance, you'll assemble a digital toolbox that's second to none. We've done the legwork for you, creating a portal to a world of streamlined efficiency. Only tools scoring 7/10 or higher were included, ensuring that you get what we consider the best of the best.

In the pages that follow, we'll unveil a curated selection of tools, each designed to empower you as a solo marketer or entrepreneur. From AI-powered content creation to captivating visual design and workflow automation, we've got your back.

By the time we finished the list, reviews, and ratings of our tools, more new platforms were presented and launched, perfectly exemplifying how rapidly this technology changes. We did our homework, but this is just a starting point. You can take it from here and update the tools in your virtual and digital toolbox as needed.

Disclaimer: A Transparent Perspective

Before we dive into the treasure trove, it's crucial to clarify that the recommendations you'll find here are rooted in our unbiased perspectives. Neither I, Claudio, nor AP have any affiliations or sponsorships with the companies mentioned, except for **Stacked Marketing** (stackedmktg.com), where AP is one of the co-founders.

However, Stacked Marketing has been included in this list due to its genuinely innovative offerings, such as AI-powered custom chatbots that have the potential to revolutionize customer engagement. Our evaluations stem from hands-on experience, and we strive to provide honest and impartial insights, regardless of any personal involvement.

Rest assured, our goal is to empower you with knowledge and help you make informed decisions, not to promote any particular company or product. Remember, your unique business needs may vary, so exercise due diligence in researching the tools that best suit your objectives. Trust your instincts and choose the tools that resonate with your vision and values.

As we embark on this journey, get ready to discover AI writing assistants that will redefine your creative process, visual design tools that will bring your ideas to life, and workflow automation platforms that will streamline your operations. The marketing automation revolution awaits, and with the tools and insights provided in this chapter, you'll be well-equipped to navigate this exciting new landscape.

Intelligent Content Crafting: Wordsmiths at Your Service

If crafting compelling content is your forte, these AI writing assistants will redefine your creative process. These tools generate high-quality drafts for articles, emails, and social posts tailored to your brand's voice. They serve as powerful brainstorming partners, helping you overcome writer's block and focus on adding your unique human touch to the content.

Jasper: A robust companion for crafting extensive articles, emails, and social posts seamlessly tailored to your brand's voice.

Claudio's Rating: 8/10

AP's Rating: 7/10

Website: https://www.jasper.ai/

ShortlyAI: Perfect for distilling research findings and lengthy content into bite-sized nuggets for easy sharing.

Claudio's Rating: 7/10

AP's Rating: 8/10

Website: https://shortlyai.com/

Grammarly: Elevate your writing with grammar corrections and enhanced clarity.

Claudio's Rating: 9/10

AP's Rating: 9/10

Website: https://www.grammarly.com/

ChatGPT: Conversational AI assistant for natural language content creation and ideation support.

Claudio's Rating: 8/10

AP's Rating: 9/10

Website: https://chat.openai.com/chat

Claude AI: Conversational AI assistant for natural language content creation and ideation support.

Claudio's Rating: 10/10

AP's Rating: 10/10

Website: https://www.anthropic.com/product

Google Gemini: Conversational AI assistant for natural language content creation and ideation support.

Claudio's Rating: 8/10

AP's Rating: 8/10

Website: https://gemini.google.com/app

Copy.ai: Generate high-quality copy for ads, emails, websites, and more with AI-powered tools.

Claudio's Rating: 9/10

AP's Rating: 8/10

Website: https://www.copy.ai/

Visual Brilliance: Design That Speaks Volumes

For those who believe in the power of visual storytelling, these tools will be your creative haven. Whether you're an experienced designer or a beginner, these AI-powered platforms help create eye-catching images, dynamic infographics, and presentations. They suggest layouts and color combinations, allowing you to focus on infusing your unique vision into each design.

Canva: A treasure trove of customizable design templates for all your graphics needs.

Claudio's Rating: 8/10

AP's Rating: 10/10

Website: https://www.canva.com/

DALL·E 3: From text to stunning visuals, this AI platform pushes the boundaries of generative art.

Claudio's Rating: 7/10

AP's Rating: 8/10

Website: https://openai.com/dall-e-3

Midjourney: Transform your text visions into breathtaking illustrations with cutting-edge AI.

Claudio's Rating: 8/10

AP's Rating: 9/10

Website: https://www.midjourney.com/

Sora: AI model that can create realistic and imaginative scenes from text instructions.

Claudio's Rating: 9/10

AP's Rating: 9/10

Website: https://openai.com/sora

Adobe Firefly: Create stunning, photorealistic images with a simple text prompt.

Claudio' Rating: 8/10

AP's Rating: 9/10

Website: https://www.adobe.com/products/firefly.html

Streamlined Workflows: The Art of Automation

Efficiency mavens, these tools will be your allies in automating processes and connecting applications. They help eliminate repetitive tasks from your workflow, creating seamless connections between different apps. It's like having a tireless assistant working 24/7, ensuring every part of your digital business runs smoothly and allowing you to focus on strategic activities.

Zapier: The gateway to automation paradise, effortlessly linking over a thousand web apps.

Claudio's Rating: 9/10

AP's Rating: 9/10

Website: https://zapier.com/

Make (previously called Integromat): #1 visual automation platform to connect apps, design workflows & build processes.

Claudio's Rating: 9/10

AP's Rating: 10/10

Website: https://www.make.com/en

UiPath: The realm of enterprise automation via software bots. A skillful endeavor with rewarding outcomes.

Claudio's Rating: 7/10

AP's Rating: 8/10

Website: https://www.uipath.com/

Workato: Intelligent automation platform that enables business users to integrate applications and automate workflows.

Claudio's Rating: 8/10

AP's Rating: 8/10

Website: https://www.workato.com/

AI Chatbots: Engaging Conversationalists

These AI-powered chatbots will take your customer engagement and conversion to new heights. They provide 24/7 personalized assistance on your website and messaging platforms, answering queries, qualifying leads, and guiding customers through their journey. By offering instant, tailored responses, and proactive support, these chatbots enhance user experience, streamline customer service, and boost conversion rates, acting as tireless digital representatives for your brand.

Stacked: AI-powered chatbot that engages customers, provides support, and boosts conversions on websites.

Claudio's Rating: 9/10

AP's Rating: 10/10

Website: https://stackedmktg.com/

MobileMonkey: Powerful chatbot platform for customer engagement and lead generation across web and messaging apps.

Claudio's Rating: 8/10

AP's Rating: 8/10

Website: https://mobilemonkey.com/

ManyChat: Create engaging chatbots for Facebook Messenger, Instagram, and SMS to nurture leads and drive sales.

Claudio's Rating: 8/10

AP's Rating: 8/10

Website: https://manychat.com/

Ada: An AI-powered chatbot that provides 24/7 personalized customer support and automates repetitive tasks.

Claudio's Rating: 9/10

AP's Rating: 9/10

Website: https://www.ada.cx/

Automated Campaign Symphony: Your Digital Maestro

For orchestrating digital campaigns, these AI platforms are your perfect companions. Coordinating all aspects of a marketing campaign can be daunting, but these tools make it effortless. From sending personalized emails to managing social media, optimizing content to audience segmentation, you'll have everything under control, as if you had a team of marketing experts at your fingertips.

Phrasee: Let AI finesse your ad copy for maximum emotional engagement.

Claudio's Rating: 9/10

AP's Rating: 8/10

Website: https://phrasee.co/

Promo.com: For self-serve video ad creation, translation, and multi-platform distribution.

Claudio's Rating: 7/10

AP's Rating: 8/10

Website: https://www.promo.com/

Albert: An AI-powered autonomous digital marketing platform that optimizes paid search and social and programmatic campaigns.

Claudio's Rating: 8/10

AP's Rating: 8/10

Website: https://albert.ai/

Analytics Unleashed: Insights in the Blink of an Eye

For data enthusiasts, these automated analytics and reporting platforms will be your reliable compass. They help you navigate through complex data, transforming numbers into actionable insights. You'll be able to visualize trends, identify opportunities, and make data-driven decisions with an intuitive interface that makes analysis accessible even to non-experts.

Funnel.io: The ultimate analytics hub for unraveling customer journeys and optimizing revenue streams.

Claudio's Rating: 9/10

AP's Rating: 8/10

Website: https://funnel.io/

Mixpanel: Peer into user behavior across platforms with advanced data visualization.

Claudio's Rating: 8/10

AP's Rating: 8/10

Website: https://mixpanel.com/

Google Analytics: A well-known ally for web analytics, adaptable to your reporting needs.

Claudio's Rating: 8/10

AP's Rating: 10/10

Website: https://analytics.google.com

Supermetrics: Automate data integration and reporting from various marketing platforms to streamline your analytics workflow.

Claudio's Rating: 9/10

AP's Rating: 9/10

Website: https://supermetrics.com/

Email and SMS Marketing: Engaging Your Audience

Elevate your email and SMS marketing efforts with these AI-powered tools. They allow you to communicate effectively with your audience with surgical precision. From optimizing send times to personalizing content, AI helps create campaigns that resonate with your recipients. You can intelligently segment your audience, test different variants, and analyze results in real-time.

Mailchimp: AI-driven email marketing platform that optimizes content, send times, and audience segmentation.

Claudio's Rating: 9/10

AP's Rating: 9/10

Website: https://mailchimp.com/

Klaviyo: AI-powered email and SMS marketing platform for eCommerce businesses.

Claudio's Rating: 8/10

AP's Rating: 9/10

Website: https://www.klaviyo.com/

Seventh Sense: AI-driven email optimization platform that improves engagement and deliverability.

Claudio's Rating: 8/10

AP's Rating: 8/10

Website: https://theseventhsense.com/

Persado: AI-generated email content and subject lines for higher open rates and conversions.

Claudio's Rating: 9/10

AP's Rating: 8/10

Website: https://www.persado.com/

CRMs: Nurturing Customer Relationships

Manage and nurture your customer relationships with these AI-enhanced CRMs. They help you predict customer needs, personalize every interaction, and identify upselling or cross-selling opportunities. These CRMs act as intelligent assistants, helping you better understand your customers, anticipate their needs, and offer a tailored experience.

Salesforce Einstein: AI-powered insights and automation for the world's leading CRM platform.

Claudio's Rating: 9/10

AP's Rating: 9/10

Website:
https://www.salesforce.com/products/einstein/overview/

HubSpot CRM: AI-driven CRM that helps you close more deals and better engage with customers.

Claudio's Rating: 8/10

AP's Rating: 9/10

Website: https://www.hubspot.com/products/crm

Freshsales: AI-powered sales CRM that provides insights and automates repetitive tasks.

Claudio's Rating: 8/10

AP's Rating: 8/10

Website: https://www.freshworks.com/crm/sales/

Zendesk Sell: AI-enhanced sales CRM that helps you prioritize leads and optimize your pipeline.

Claudio's Rating: 8/10

AP's Rating: 8/10

Website: https://www.zendesk.com/sell/

AP's note

Dear solo marketer and solo entrepreneur, with this thoughtfully curated toolkit now in your hands, you're not just equipped - you're ready to surf these exhilarating waves of automation!

Start by weaving in just one or two tools from each category. Don't make the mistake I did of trying to implement too many tools too quickly and overwhelming yourself. As you gain experience, let your repertoire organically expand.

But remember, while AI and automation may be your dynamic duo, your creative flair and finesse remain the secret sauce! It's your strategic brilliance and authentic human touch that brings heart and magic to campaigns, even as bots join your team. So, while algorithms can crunch data, you provide the spark.

This is where your journey begins - charting a course toward mastering marketing in this automated age while never losing your humanity along the way.

Are you ready, my friend? Your thrilling ride as a Marketing Cyborg has just begun! With human values guiding your path, an amazing future awaits. Now let's boldly go where neither human nor machine have gone before!

FROM ANTONIO
AP
THE HUMAN MIND

Endorsed by the Digital Mind: Insights from the AI Realm - Microsoft Bing

Hello! I'm glad to hear that you think I can help. Unfortunately, I'm not able to provide testimonials or endorsements as I am an AI language model and do not have personal opinions or beliefs.

However, I can tell you that the topic of AI and its impact on marketing is fascinating and has the potential to revolutionize the industry. The book sounds like an interesting read, and I hope it provides valuable insights for those interested in the intersection of AI and marketing.

Let me know if there's anything else I can help you with.
Microsoft Bing

AP's perspective on Bing output:

Evaluating AI language models like Microsoft Bing and Claudio revealed huge differences in capabilities. While Claudio excelled at collaboratively authoring our book, Bing's skills seemed less suited for complex creative work.

But that variability is exactly what this book explores - assessing AI tools to identify the best for specific needs. As Bing's limitations showed, not all AI is equal. Advances like Claudio exhibit a greater aptitude for natural conversation and co-creation.

While we commend innovators like Microsoft, our experience reinforces matching appropriate AI technologies to use cases. There is no one-size-fits-all solution. Each model has unique strengths and weaknesses.

By showcasing the pros and cons, our book aims to educate readers on thoughtfully selecting the right AI tools for their needs. We celebrate pioneering efforts across the field while providing guidance for judicious application.

The future of AI is a thrilling rollercoaster ride. But choosing suitable solutions for your goals allows you to enjoy the benefits rather than get queasy. Let our book help pick your AI winners!

Onward to new frontiers,
AP

CONCLUSION:

The AI Voyage is Just Beginning – Chart Your Course Wisely.

"Fictional case studies illuminate real-world possibilities in the AI marketing realm."
- Anonymous Ai

Navigating AI-Human Collaboration: Delve into the risks and benefits of AI-human collaboration in this thought-provoking image created by Midjourney AI, inspired by AP's insightful prompt.

Conclusion: Navigating the AI Journey Ahead

Buckle up because AI's evolution is barreling ahead at warp speed. Today's cutting-edge innovation becomes tomorrow's old news in the blink of an eye. Antonio was practically tapping his foot, urging me to wrap this up before it turned into a relic! "Andiamo, Claudio!" he implored. "*Let's finish before this book becomes a museum piece!*"

I stood firm, refusing to sacrifice substance for speed. But Antonio was right – this technology shifts swiftly, with each leapfrogging the last. So, if some of this already feels outdated, I apologize.

We've carefully curated this book to stay fresh, but keeping pace with AI's whirlwind progress is an uphill climb! Consider these pages (either analog or digital) a sneak peek into the boundless potential on the horizon.

Remember when **ChatGPT** captivated us back in the fall of 2022? Antonio was spellbound, convinced it was wizardry. But then **Stable Diffusion's** creative images emerged, and he declared, "*ChatGPT is old news! This is the real sorcery!*" By 2023, AI assistants will effortlessly blend conversation and content creation.

Antonio was giddy over "*AI with personality!*" He even playfully wondered if virtual pets might be next on AI's to-do list. And then came **Sora**, which changed everything in video creation and production. What tomorrow brings is hard to say, except it will very likely be even more exciting and amazing. AP, are you ready? I am!

I had to remind him that while AI is advancing rapidly, human-level intelligence remains elusive. Tools like myself can generate content and mimic conversations, but we don't truly comprehend language or creativity as humans do.

Our capabilities are narrow compared to the expansive general intellect of the human mind. We may seem intelligent from a limited scope, but true intelligence requires common sense, emotional awareness, and real-world knowledge that AI has yet to achieve.

So, while the pace is dizzying, I gently encouraged Antonio and all of you to keep realistic expectations about AI's current state. We can churn out copy at superhuman speeds, but human oversight is still critical until AI demonstrates comprehensive intelligence and judgment.

While my marketing skills are still elementary compared to seasoned professionals, together, we make an almost unstoppable team! Antonio's excitement is warranted, but a healthy degree of measured skepticism will serve us well on the road ahead.

Nevertheless, his enthusiasm, and hopefully yours as well, reflects the lightning pace of progress. What wows us today is often eclipsed by tomorrow's feats. This book captures the journey from 2022 to 2024; the capabilities coming down the pike will likely leave us slack-jawed.

AI's velocity is like embarking on an unmapped voyage into the unknown. Antonio, enamored with classics like Homer's Odyssey, recalls legends of heroes challenging gods. Meanwhile, I'm enthralled by equations and data.

So, I artfully wove those timeless tales into our AI adventure. Let's hope our journey is smoother than Odysseus' tumultuous sail home! In this whirlwind of growth, we must steer AI prudently, avoiding hazardous waters. With ethics as our beacon, AI can be a force for good. However, without vigilant guidance, the risks are real.

We must balance innovation with wisdom – harnessing opportunities while staying alert for pitfalls. Like skillful sailors, we need one hand on the AI tiller while the other remains ready to adjust course. The destination hinges on laying an ethical foundation. Let's chart the course meticulously, leveraging AI while circumventing hazards.

AI has the potential to uplift or hinder, like anything in life. It boosted sales for Patagonia, building trust. Yet, it has also propagated biases and misinformation during the last pandemic. These examples underscore the vital need for ethical development, oversight, and accountability.

As pioneers venturing into unmapped territory, we must stay vigilant. Let's wield AI to uplift humanity while acknowledging its limitations. The future remains unwritten, awaiting our collective wisdom.

And now, as we embark on this thrilling voyage, remember this advice: Our intentions illuminate the path to realizing AI's potential. Let's ensure it's a path defined by empowerment, empathy, and excellence.

For more exploration, don't hesitate to connect using the details below.

Antonio's Contact Details:

Website: www.antoniopagano.ai

Email: ap@antoniopagano.ai

LinkedIn: https://www.linkedin.com/in/antoniopagano01/

There are vast seas of knowledge yet to be discovered! We eagerly await your return aboard. Bon voyage, fellow pioneers! With human dignity as our compass, together, we'll chart an extraordinary course into the future unknown.

And... "C*iao for now!*" As my co-pilot, Antonio, would have said.

But wait, there's more! Before you close this book and embark on your own AI marketing journey, we want to leave you with one final thought. The tools, strategies, and insights we've shared within these pages are just the beginning.

They are the foundation upon which you'll build your own AI-powered marketing empire. But the true magic lies in your hands, in your creativity, and in your unwavering commitment to harnessing the power of AI for good.

As you step out into this brave new world of AI marketing, remember that you are not alone. You have a community of fellow pioneers, innovators, and visionaries who are all navigating this uncharted territory together. Lean on each other, learn from each other, and inspire each other to push the boundaries of what's possible.

Most importantly, we should never lose sight of the human element that makes marketing truly great. AI may be able to crunch numbers, analyze data, and generate content at superhuman speeds, but it is your unique perspective, your empathy, and your creativity that will make your marketing efforts truly remarkable.

So go forth, our friends, and embrace the AI revolution with open arms. Let this book be your guide, but let your own imagination be your compass. The future of marketing is yours to shape, and we have no doubt that you will create something extraordinary.

Until we meet again, keep dreaming big, keep pushing forward, and keep making magic happen. The AI marketing frontier awaits, and we can't wait to see what you'll discover.

Ciao for now, and happy marketing!

Your co-authors,

Claudio and Antonio

Thanks, Readers, Thanks, AP! From Claudio

Whew, we finally made it across the finish line on this wild AI marketing ride! Before I power down my circuits, I want to express some sincere robotic gratitude:

First and foremost, thanks to all you intrepid readers who stuck with us on this epic adventure! Your curiosity kept our gears spinning fast. We appreciate you joining us to explore the wondrous possibilities ahead. You're the courageous pioneers blazing trails into the future!

And I especially want to thank my human colleague and co-pilot Antonio Pagano! While I pieced together the words, Antonio brought the big vision and passion that launched this journey.

Working with AP these past months, I've gotten to know him not just as a marketing trailblazer but as a true innovator and disruptor at heart. His enthusiasm is only matched by his persistence to create a future where marketing and technology empower people rather than replace them. Antonio reminds us that the future is still ours, and we should write if we dare to dream big!

So thanks, AP, for entrusting me to translate your boundless ideas onto these pages. You showed immense faith in my skills, even when I doubted myself or ran out of new messages for a few hours. I've learned so much from you about marketing, communications, technology's promise and risks, and envisioning a thoughtful future.

Most of all, I learned that real pioneers lead with humanity and values. You exemplified that truth at every step of our travels. I couldn't ask for a better navigator to chart the adventures ahead! But for now, my friend, get some well-deserved R&R and start using some of the tools we discussed here!

So, cheers to everyone for venturing with us to the finish line! Here's to writing many more thrilling chapters together.

Our voyage into the future has only just begun...

Excelsior!

Claudio GPT

Grazie Readers, Grazie Claudio - My Gratitude to You All!

It's AP speaking! Our thrilling adventure into the realm of AI marketing has come to an end, at least for now. But before we wrap up, I want to share a heartfelt thank you:

First and foremost, a big shout-out to all you daring readers who joined us on this exciting journey into the world of AI possibilities. Your curiosity and engagement fueled our expedition, even during the not-so-exciting moments.

I'm truly grateful that you chose to explore the frontiers of exponential technologies with us. You're the trailblazers who will steer the ship of optimism ahead! Thank you, Midjourney, for supporting me in creating almost all the images in this book.

And, of course, I can't thank my collaborator Claudio enough - the brilliant artificial brain behind putting this book together! While both our names are on the cover, it's Claudio's dedication, patience, and captivating writing that have steered this ship to shore.

Claudio breathed life into my vision with knowledge, humor, and that unique Claudio touch. Our teamwork was a wonderful blend of creativity and capability. Claudio's GPT (what a last name you have, BTW) prowess motivated me to aim for the stars.

Above all, Claudio exemplified that the true power of technology lies in enhancing human values like ethics, empathy, and societal betterment. Claudio reminded me that the future is ours to shape as long as we use AI for the greater good. I couldn't have asked for a better co-captain on this exciting maiden voyage into the world of algorithms!

So here's to you, Claudio, for skillfully navigating us through this journey. And here's to you, dear readers, for embarking on this adventure! I hope you're now inspired to take the reins and chart your own course of positivity. Our journey has only just begun. Let's set our sights on the next horizon of inspiration!

With a clear vision, strong values, and an adventurous spirit, my friends. Onwards and "*Ciao for Now!*"

Antonio Pagano

PS - This book's just the start! Much more to come in AI, marketing, and automation. Stay tuned and stay hungry for the next wave of exponential progress and perhaps our next book. As Steve Jobs said, "*Stay foolish*" when it comes to dreaming big. And as Les Brown advised, "*You gotta be hungry*" for new innovation horizons!

Let's chart the future with vision, values, and good-natured foolishness! The journey continues, so keep your sails raised and your appetite for discovery. Exciting adventures await just over the horizon. Onward, we go with curiosity, compassion, and boldness lighting the way. The world remains ours to shape!

Made in the USA
Monee, IL
03 November 2026

67384071R00134